"In *The Global Governance of Genetic Resources* Florian Rabitz deals with utterly pressing issues in international politics today, while making a significant contribution to the scholarly literature on institutional complexity and institutional change. The book intelligently engages with contemporary problems of our time; it promises a lasting impact on our ability to understand why and how international institutions evolve."

Prof. Dr. Cristiane Lucena, Institute for International
Relations, University of São Paulo

"Academic interest in various forms of institutional change that are provoked by overlapping international institutions has increased sharply in recent years. Academic journals have been accommodating a vibrant debate among IR scholars. Florian Rabitz enriches this debate by having written one of the first books that tackles this important phenomenon. Rabitz develops a novel and compelling theoretical account of different forms of institutional change unfolding within regime complexes, which he applies carefully to the global governance of genetic resources. This book is not to be missed by those interested in the implications of institutional complexity on global governance from a theoretical angle, much less by the ones keen to know more about the global governance of genetic resources from an empirical point of view."

Dr. Benjamin Faude, WZB Berlin Social Science
Center, Research Unit "Global Governance"

"Crossing disciplinary boundaries, Florian Rabitz explores how and why international institutions governing genetic resources evolve. He convincingly argues that patterns of interests and interdependence among states explains changes at the Convention on Biological Diversity, the World Trade Organization, the Food and Agriculture Organization, the World Health Organization, and other institutions governing genetic resources. Combining theoretical rigor and empirical insight, *The Global Governance of Genetic Resources* offers a dynamical representation of international negotiations. This book will be of interest to political scientists, legal scholars, economists, and professionals working on global governance."

Prof. Dr. Jean-Frédéric Morin, Canada Research Chair
in International Political Economy, University of Laval

W0113614

The Global Governance of Genetic Resources

Multi-institutional governance architectures are increasingly common in world politics, yet how do they evolve over time? This book develops a fresh conceptual approach by distinguishing two main types of institutional change and proposing the strategic context within which governments make decisions regarding international cooperation as the main driving factor. Applying this theoretical framework to the case of genetic resources, it shows how the scope for change has persistently been circumscribed by asymmetries in the global biotechnology sector. Taking a broad view of the underlying technological, legal and economic factors, the book analyzes the formation of international regimes linking access to genetic resources to the fair and equitable sharing of the benefits arising out of their utilization. Covering negotiations in the areas of seeds, intellectual property rights, pandemic influenza viruses and marine genetic resources, the author shows how governments have persistently faced the problem of ensuring cooperation among actors with widely differing interests. This led them to opt for a strategy of institutional layering, whereby new international instruments are gradually built upon pre-existing ones. In addition to giving a comprehensive overview of the international governance of Access and Benefit-sharing within the wider context of modern biotechnology, the argument developed here enables a new perspective for studying institutional change in multi-institutional governance architectures.

Florian Rabitz is a postdoctoral researcher at Kaunas University of Technology. He has previously taught at the University of São Paulo and holds a PhD in political science from the Free University of Brussels. His research focuses on international institutions and institutional change in global environmental governance. Previous work has appeared in outlets such as *Futures*, the *Journal of International Relations and Development*, *Third World Quarterly* and the *Journal of European Public Policy*.

Global Environmental Governance
Series Editors: John J. Kirton, Miranda Schreurs

Global Environmental Governance addresses the new generation of twenty-first century environmental problems and the challenges they pose for management and governance at the local, national, and global levels. Centered on the relationships among environmental change, economic forces, and political governance, the series explores the role of international institutions and instruments, national and sub-federal governments, private sector firms, scientists, and civil society and provides a comprehensive body of progressive analyses on one of the world's most contentious international issues.

Most recent titles

The Global Governance of Climate Change: G7, G20, and UN Leadership
John J. Kirton and Ella Kokotsis

Moving Health Sovereignty in Africa: Disease, Governance, Climate Change
Edited by John J. Kirton, Andrew F. Cooper, Franklyn Lisk and Hany Besada

Africa's Health Challenges: Sovereignty, Mobility of People and Healthcare Governance
Edited by Andrew F. Cooper, John J. Kirton, Franklyn Lisk and Hany Besada

Corporate Responses to EU Emissions Trading: Resistance, Innovation or Responsibility?
Jon Birger Skjærseth and Per Ove Eikeland

The Global Governance of Genetic Resources

Institutional Change and Structural Constraints

Florian Rabitz

Routledge
Taylor & Francis Group

LONDON AND NEW YORK

First published 2017 by Routledge

2 Park Square, Milton Park, Abingdon, Oxon OX14 4RN
605 Third Avenue, New York, NY 10017

Routledge is an imprint of the Taylor & Francis Group, an informa business

First issued in paperback 2021

Publisher's Note

The publisher has gone to great lengths to ensure the quality of this reprint but points out that some imperfections in the original copies may be apparent.

British Library Cataloguing-in-Publication Data
A catalogue record for this book is available from the British Library

Library of Congress Cataloging-in-Publication Data
Names: Rabitz, Florian.
Title: The global governance of genetic resources: institutional change and structural constraints / Florian Rabitz.
Description: Abingdon, Oxon; New York, NY: Routledge, 2017. | Series: Global environmental governance | Includes bibliographical references and index.
Identifiers: LCCN 2016048055 | ISBN 9781138281110 (hardback) | ISBN 9781315271316 (e-book)
Subjects: LCSH: Germplasm resources. | Germplasm resources, Animal. | Germplasm resources, Plant. | Animal genetics. | Plant genetics. | Biotechnology.
Classification: LCC QH430 .R33 2017 | DDC 333.95/34—dc23
LC record available at https://lccn.loc.gov/2016048055

ISBN: 978-1-138-28111-0 (hbk)
ISBN: 978-1-03-217914-8 (pbk)
DOI: 10.4324/9781315271316

Typeset in Times New Roman
by codeMantra

Contents

Figures and tables

Figures

Tables

Acronyms

ABNJ	Areas Beyond National Jurisdiction
ABS	Access and Benefit-sharing
ABSWG	Ad Hoc Open-ended Working Group on Access and Benefit-sharing
CBD	Convention on Biological Diversity
COP	Conference of the Parties
FAO	Food and Agriculture Organization
GISN	Global Influenza Surveillance Network
GISRS	Global Influenza Surveillance and Response System
ICP	Open-ended Informal Consultative Process
IGC	Intergovernmental Committee on Intellectual Property and Genetic Resources, Traditional Knowledge and Folklore
IGM	Intergovernmental Meeting
IHRs	International Health Regulations
ISA	International Seabed Authority
MGR	Marine Genetic Resources
PCT	Patent Cooperation Treaty
PGRFA	Plant Genetic Resources for Food and Agriculture
PIP	Pandemic Influenza Preparedness
PrepCom	Preparatory Committee Established by the UN General Assembly Resolution 69/292
SMTA	Standard Material Transfer Agreement
SPLT	Substantive Patent Law Treaty
TRIPS	Agreement on Trade-related Aspects of Intellectual Property Rights
UNCLOS	United Nations Convention on the Law of the Sea
WG-BBNJ	Ad Hoc Open-ended Informal Working Group to Study Issues Relating to the Conservation and Sustainable Use of Marine Biological Diversity Beyond Areas of National Jurisdiction
WHO	World Health Organization
WIPO	World Intellectual Property Organization
WTO	World Trade Organization

Acknowledgements

This book is based on a dissertation which I completed at the Vrije Universiteit Brussel in 2014. First and foremost, I would like to thank my former supervisor, Sebastian Oberthür, whom I could always rely on for advice and support. I would also like to thank Matthias Buck, Benjamin Faude, Harri Kalimo, Joachim Koops, Jean-Frédéric Morin, Amandine Orsini and Justyna Pozarowska, all of whom contributed in different ways to the final text, as well as all the (other) wonderful people at the Institute for European Studies, Vrije Universiteit Brussel, and at the Institute for International Relations of the University of São Paulo, where I finalized the manuscript.

Florian Rabitz
Kaunas, February 2017

1 Introduction

Probably every social, political, environmental or economic issue falls within the ambit of some international treaty, organization or partnership, some declaration, code of conduct or advisory body. From macroeconomic policy coordination, cross-border investments and intellectual property rights, labor standards, election monitoring and satellite orbits to infectious diseases, the trade in endangered species, hazardous chemicals and ambient air pollution, all politics is global.[1] The number of international institutions is arguably growing faster than the number of problems they are supposed to address.

Even more remarkable than the fact that there are very few facets of daily life that are not in some way touched upon by some international institution is the fact that most are touched upon by more than one. The bus we take to work might be produced according to international standards regulating flue gas emissions, acquired in line with international agreements on government procurement and driven on a blend of biodiesel and petrodiesel mandated as part of the public authorities' larger attempts at implementing their commitments to reduce national greenhouse gas emissions. Traveling abroad by plane, we might be subject to an international visa waiver scheme, our ticket price could include surcharges for greenhouse gas emissions from aviation, airport security will be in line with international minimum standards and, in case our baggage gets lost, we are entitled to an internationally agreed-upon minimum amount of compensation.

The existence of multiple international institutions with different degrees of responsibility for a wider issue area is, today, the norm rather than the exception. Consider the field of arms control. Some international treaties regulate specific types of weapons systems: anti-personnel mines are regulated under the 1997 Ottawa Treaty and a protocol to the 1981 Convention on Certain Conventional Weapons, the latter also covering laser and incendiary weapons. Chemical and biological agents respectively fall under the 1993 Chemical Weapons Convention and the 1972 Biological and Toxin Weapons Convention. Nuclear weapons are the focus of the Threshold-, Partial- and Comprehensive Nuclear Test Ban Treaties and the Non-Proliferation Treaty and, at the bilateral and regional levels, of the New START treaty or the Pelindaba, Tlatelolco and Rarotonga treaties; weather warfare falls under the Environmental Modification Convention; cluster bombs under the Convention on Cluster Munitions; trade in different conventional weapons is addressed by the 2013 Arms Trade Treaty; missile technology falls under various non-binding agreements such as the International Code of Conduct against Ballistic Missile Proliferation or the Missile Technology Control Regime, and the use of arguably most types of weapons is indirectly regulated by a host of human rights treaties.

This example shows the institutional diversity and growing institutional density we find in the international system today: different institutions exist for different categories of weapons systems, sometimes even for different aspects of a single category; other institutions take a broader scope, applying to a broad range of different categories. Some of those institutions take the form of binding international law; others are declarative in nature. Some of the institutions are genuine international organizations (such as the Chemical Weapons Convention); others are directly executed by governments (such as the Biological and Toxin Weapons Convention). Some include inspection regimes and other forms of compliance mechanisms; others do not. Some have almost universal membership; others are restricted to the regional or bilateral level.

The topic of this book is institutional change in complex settings of multiple, overlapping international institutions. Institutional change is a perennial problem of the social sciences. As institutions can broadly be understood as predictable and persistent patterns of social behavior, "change" is initially an antithetical concept. Generations of scholars have struggled to find satisfactory explanations for institutional change. In political science, theories of institutional change have predominantly been developed within comparative politics and less so in international relations.[2] Explaining changes in wider arrays of "partially overlapping and nonhierarchical institutions governing a particular issue area"[3] instead of in single institutions is a different beast entirely. What types of change exist in such larger, multi-institutional regime complexes, and how do we explain them? Several authors have grappled with this question. Some argue that change is driven by interest differentials: once dissatisfied actors cannot reform a given institution, they choose to create alternatives within the same issue area, leading to the emergence of overlaps.[4] Others focus on how reformist actors attempt to shift the operations of existing institutions into new directions, possibly extending their regulatory scope into new areas.[5] Still others analyze those drivers that, over time, lead regime complexes to evolve from competitive inter-institutional relations towards an established division of labor in which each elemental institution fulfills a specialized role, and interferences among them are minimal.[6] Most authors generally acknowledge that, to some extent, institutional change in regime complexes is path dependent.[7] Considering the ambiguity of the latter concept,[8] however, this raises more questions than it answers: what are the causes of path dependence, should we differentiate between different *types* of path dependence and how does path dependence relate to concepts such as "stability" that are also frequently invoked in regards to regime complexes?[9]

The questions of what exactly constitutes "institutional change" and how to demarcate different types of change are always somewhat in the eye of the beholder.[10] As with the specialized literature on regime complexity, the literature on institutional change, if such a literature even exists in the singular, is cluttered with concepts, taxonomies and ambiguities, most of which I will be unable to resolve in this text. The theoretical objective is more pragmatic: to elaborate the foundations of a parsimonious theory that can explain institutional changes in regime complexes in a satisfactory manner—that is, explaining a broad range of empirical phenomena with as little theoretical assumptions as possible.[11]

In a nutshell, my argument is this: institutional changes in regime complexes are of two types. Club cooperation refers to the phenomenon that states, unable to achieve their respective policy objectives within wider, multilateral settings, band together to form smaller cooperative arrangements. By limiting the variance of interests among the participating actors, they are able to agree on deeper (or simply different) types of rules than would be possible with a broader membership. This phenomenon is well known. It can be observed in numerous fields and has been used in contemporary approaches to institutional change in regime complexes.[12] Its counterpart, what I refer to as institutional layering, is a concept with a long tradition in comparative politics yet has so far not been applied to the field under consideration here. Institutional layering broadly refers to incremental changes whereby actors build upon existing institutions in order to develop them into their preferred direction. Rather than restricting institutional membership for limiting the variance of interests as with club cooperation, reformist actors employing institutional layering attempt to maintain broad, inclusive cooperation. To ensure the participation of those actors that are oriented towards the institutional status quo, the extent of institutional change needs to be restricted, or concessions in other areas must be offered through issue linkage.

Empirically, this book limits itself to this second type of institutional change, which has, so far, not been thoroughly studied.[13] I thus do not subject my theoretical approach to a comprehensive test, which would require including cases where institutional change takes the form of club cooperation. I return to the question of generalization in Chapter 9. The theory I develop here revolves around situation structure: actors' relative capacities to supply collective goods and to generate transnational negative externalities. I understand situation structure from two sides: the specific patterns of interdependence it constitutes and the interests to which it gives rise. The participation of an actor causing large negative externalities, for instance, is indispensable for cooperative attempts at limiting or reducing the particular externality. This is a question of (asymmetrical) interdependence. At the same time, an actor who generates negative externalities profits from doing so while the costs are borne out by others. He will, accordingly, prefer the institutional status quo to alternative institutional arrangements that mandate greater efforts at reducing negative externalities. This is a question of interests. Institutional layering thus occurs when *demandeurs* of institutional change have no viable outside option due to the contributions status quo-oriented actors make towards joint gains from cooperation and/or due to the costs their non-participation would generate.

I test this hypothesis against two competing explanations. The early literature on regime complexity argued that institutional change proceeds through *regime shifting*, when states (and non-state actors) seek to bring a given issue, or set of issues, under a new regulatory approach.[14] When new institutions are being brought in to govern a particular issue area, they generate overlaps and potential inconsistencies with older ones. The second, more recent explanation is the exact opposite: states have a general interest in securing institutional functionality and complementarity. *Interplay management* thus refers to attempts to avoid problematic interactions between international institutions while implementing pre-existing commitments.

I apply those theories to the genetic resources regime complex, which has been a focal object of study in the literature.[15] This complex revolves around the question of who can access and utilize biological materials from plants, microorganisms or animals and under which conditions. Within this complex, different types of international institutions, from areas such as biodiversity, intellectual property, international public health, the law of the sea and agriculture overlap with each other. Both regime shifting and interplay management have been used for explaining institutional change within this regime complex. Methodologically, I focus on regime formation within the issue area of genetic resources: the 2001 International Treaty on Plant Genetic Resources for Food and Agriculture, the 2010 Nagoya Protocol to the Convention on Biological Diversity, the 2011 Pandemic Influenza Preparedness Framework and, as a moving target, the ongoing negotiations on an international regime for marine genetic resources. I employ a mixture of congruence procedures, process tracing and counterfactual reasoning: first, I assess how far institutional outcomes are in line with what situation structure, regime shifting and interplay management would predict. Second, I use certain diagnostic pieces of evidence from the various negotiation processes: concessions, trade-offs, issue linkages and their respective timings. This allows me to adjudicate among the three hypotheses that are, in many ways, observationally equivalent. Third, as situation structure is supposed to provide a general explanation of institutional change in regime complexes, I ask the counterfactual question of "what would have happened if the *demandeurs* of institutional change had not opted for institutional layering but for club cooperation instead?" The point of this exercise is to show that, given situation structure, layering is the only viable approach to institutional change. For the emerging regime for marine genetic resources, this entails not a counterfactual but a scenario analysis instead.[16]

Why does the genetic resources regime complex change through layering? As I attempt to demonstrate, it is a result of the asymmetrical distribution of genetic raw materials as well as the scientific and technological capacities for their utilization. The complex revolves around the twin issues of access to genetic resources and the subsequent sharing of the resulting benefits, either with the original providers of the raw materials or multilaterally. For many types of benefits, this is a zero-sum game: governments are required to redistribute some of the commercial profits (or other types of gains, such as research results or ownership of intellectual property) realized by their domestic industries to third parties. Without the cooperation of those governments, benefits will not be shared: as commercial users have few incentives to do so voluntarily, their respective governments must ensure adequate measures for monitoring and compliance. Yet, they receive relatively little in return. While industries within their jurisdictions are biased against benefit-sharing yet have stakes in accessing genetic resources, the realization of access is less dependent on international cooperation than is the realization of benefit-sharing: genetic raw materials can be sourced from different countries, including *ex situ* gene banks, and their transboundary flows are difficult to monitor. Conversely, ensuring that benefits are shared fairly and equitably requires user countries to monitor the value chain through which utilization takes place and impose remedies for situations in which industry does not comply with its obligations. Monitoring may impose costs on industry, and enforcement measures, depending on their specific design, may hamper innovation in biotechnology, notably in the non-commercial, public sector.

The basic problem is thus that those countries where utilization takes place have no reason to share the benefits they generate and have little to gain from international cooperation. Those countries that do not utilize genetic resources have reasons to ensure benefit-sharing by creating strong international institutions yet have little leverage for prodding the former into cooperation. Institutional layering thus seeks to strengthen and broaden the scope of existing international institutions to oblige user countries to take appropriate domestic measures; yet in order to ensure the latter's continuing participation, ambitions must be reduced and concessions offered in return.

The organization of this book

Chapter 2 will delve into the contemporary debate on regime complexity, specific types of multi-institutional architectures that have been identified in fields ranging from maritime piracy, climate change and arctic governance over refugee governance, food security and energy to hazardous waste and forest management.[17] I distinguish two types of institutional change in regime complexes: club cooperation (limiting interest differentials among cooperating states by creating small-group settings, thus allowing for deeper rules to be agreed on than would be possible multilaterally) and institutional layering (creating novel rules on top of older institutions in order to ensure the participation of status quo-oriented actors). Empirically, this book deals with the second type of change. I develop a theoretical approach based on the concept of "situation structure": the patterns of interests and interdependence under which states make choices about cooperation.[18] I also draw on the existing literature to identify two alternative approaches to institutional change in regime complexes: interplay management and regime shifting.[19]

Chapter 3 turns to the role of genetic resources within the broader nexus of technological, legal and economic changes taking place since roughly three decades. I address the emergence of biotechnology since the 1970s as well as the ways in which genetic resources from plants, viruses and marine organisms are being used in sectors such as pharmaceuticals, agriculture and even cosmetics.

The emergence of modern biotechnology went hand in hand with legal changes in intellectual property law. I trace the evolution of patent law in those jurisdictions that dominate global biotechnology: the US and Europe. I show how, through a series of landmark cases, the scope of patentable subject matter in micro-organisms and plant varieties has significantly expanded throughout the last 30 years. Those changes in domestic patent law are mirrored on the international level, with treaties such as the World Trade Organization's (WTO) Agreement on Trade-related Aspects of Intellectual Property Rights (TRIPS) and the International Convention for the Protection of New Varieties of Plants (UPOV in the French acronym) setting mandatory minimum standards for their members' domestic legislation. Chapter 3 also links technological and legal changes to the emergence of the modern biotech industry. I discuss the emergence of commercial biotechnology in the early 1990s before turning to the historically unprecedented wave of mergers and acquisitions that, over the course of a few years, led to the creation of "global players" such as Monsanto, Syngenta, Pioneer Dupont, GlaxoSmithKline and so forth. Genetic resources play only a minor role in those players' portfolios. Yet, the sheer scope of their commercial operations implies that they are potentially large sources of benefits to be subsequently shared.

Chapter 4 turns to property regimes over genetic resources and their historical co-evolution: the common heritage of mankind, private property rights and national sovereignty. Those regimes determine who has the rights to own, control or use genetic resources and under which conditions. Tensions among those three regimes are at the core of the political conflicts in the global governance of genetic resource and have given rise to inconsistencies both within the relevant international regimes and between them. The role of property regimes is particularly important in regards to the problem of "biopiracy," broadly referring to the illegitimate, or even illegal, appropriation of genetic resources through intellectual property rights.[20] Despite the significant problems in defining the term, concerns over biopiracy have been the primary factor driving the behavior of developing countries in the relevant international negotiation processes.

The empirical core of this book is found in Chapters 5 to 8. Here, I test how well situation structure, interplay management and regime shifting can explain the formation of four different regimes within the broader regime complex. Methodologically, I use a combination of process tracing, congruence procedures and counterfactual analysis for studying regime formation process across multiple international venues. Drawing mainly on official records, governmental submissions and reports by third-party observers, I attempt to test the relevance of each hypothesis by looking at linkages, changes in bargaining positions and their respective timing, within the various negotiation streams.

Chapter 5 addresses the process towards a regime for Access to and Benefit-sharing (ABS) from plant genetic resources for food and agriculture. This process started in the early 1980s with the International Undertaking on Plant Genetic Resources, continued with its amendments in the late 1980s and early 1990s and finished with the conclusion of the International Treaty on Plant Genetic Resources on Food and Agriculture ("Seed Treaty") in 2001. The Seed Treaty was negotiated among concerns that access to plant germplasm for the breeding of new varieties of plants would be hampered both by the growing number of intellectual property claims on plants and their genetic parts and components, as well as the tendency of provider countries to restrict access to genetic resources under their control in the hopes of obtaining large pay-offs from their utilization abroad—a problem that commenced with the conclusion of the Convention on Biological Diversity (CBD) in 1992.

Chapter 6 turns to the negotiations on an international regime to address biopiracy. This process crossed multiple international forums, including the WTO and the World Intellectual Property Organization (WIPO). As the CBD requires its contracting parties to ensure that the benefits arising out of the utilization of genetic resources are to be shared fairly and equitably, and that initial access requires prior authorization by the provider country, this process was about getting user countries to implement domestic compliance measures. I follow the multi-forum negotiations until 2010, when the Nagoya Protocol on Access and Benefit-sharing to the CBD was concluded.

Chapter 7 covers pandemic influenza viruses. For decades, viral materials had been shared freely among member states of the World Health Organization (WHO), various laboratories and commercial vaccine manufacturers. In 2007, this system ground to a halt when Indonesia stopped cooperating with the WHO, pointing out that its provision of viral materials through the WHO network is not matched by benefits, such as vaccines, flowing back. Instead, patents on vaccines create price barriers for developing countries, and Advance Purchase Agreements allow industrialized countries to absorb the entire global supply of vaccines, in case an influenza pandemic strikes. This led to a negotiation process that concluded in 2011 with the adoption of the Pandemic Influenza Preparedness Framework: an ABS regime that links multilateral virus-sharing with the sharing of benefits such as vaccines. The key dispute in this process is: are viruses genetic resources that fall under the sovereignty of nation states and do provider countries accordingly have the right to deny access despite the fact that manufacturers urgently need these viruses to produce the largest amount of vaccines possible within a short period of time?

Chapter 8 turns to the ongoing negotiation process on an ABS regime for marine genetic resources in Areas Beyond National Jurisdiction. Those negotiations have been ongoing since 2005 and are embedded within the wider objective of creating a global agreement for protecting marine biodiversity. The increasing use of marine genetic resources in so-called "blue biotechnology" is a relatively novel phenomenon, yet negotiations face several challenges, not the least of which is the legal division of the oceans into distinct zones, with different rights and obligations pertaining to each.

The analysis of those four cases shows that institutional changes in the genetic resources regime complex are largely explicable in terms of underlying patterns of interdependence and interests. Each of the competing hypotheses, interplay management and regime shifting renders partial explanations. Interplay management is strong in the case of marine genetic resources, whereas regime shifting excels for viral genetic resources. Yet their overall explanatory power is somewhat limited. In Chapter 9, I thus turn to the question of how situation structure might provide an alternative paradigm for explaining institutional change in regime complexes. I conclude this book with some observations on the future of ABS governance.

Notes

1 Drezner (2007).
2 For exceptions, see for instance Fioretos (2011); Hanrieder (2014, 2015).
3 Raustiala and Victor (2004), 279.
4 Colgan et al. (2012); Urpelainen and van de Graaf (2015).
5 Helfer (2009); Muzaka (2010); Rabitz (2015).
6 Gehring and Faude (2014); Morin and Orsini (2014).
7 i.e., Raustiala and Victor (2004); Oberthür and Pozarowska (2013); Zelli and van Asselt (2013).
8 Mahoney (2000); Pierson (2004); Capano (2009); Beyer (2010).
9 Oberthür and Stokke (2011), 322.
10 Capano (2009).
11 King et al. (1994), 29–31.
12 Colgan et al. (2012).
13 Exceptions are Helfer (2009); Muzaka (2010). A condensed version of the argument of this book has appeared in Rabitz (2015).
14 Helfer (2009).
15 Raustiala and Victor (2004); Schaffrin et al. (2006); Andersen (2008); Oberthür and Pozarowska (2013); Morin and Orsini (2014); Rabitz (2015).
16 Counterfactual reasoning and scenario analysis take a similar approach. However, while counterfactuals concern things that did not happen in the past, scenario analysis analyzes things that could happen in the future.
17 Struett et al. (2013); Keohane and Victor (2011); Stokke (2013); Betts (2009); Magulis (2013); Colgan et al.(2012); Marcoux and Urpelainen (2012); Gupta et al. (2015).
18 Zürn (1992); Mitchell and Keilbach (2001).
19 Helfer (2009); Muzaka (2010); Oberthür and Pozarowska (2013); Gehring and Faude (2014).
20 Mgbeoji (2006); Robinson (2010); Smallman (2013); Rabitz (2015).

2 Regime complexity and institutional change

The number of international institutions has skyrocketed in recent decades. The International Committee of the Red Cross, for instance, lists 27 treaties in the field of international humanitarian law.[1] The World Trade Organization presently counts 280 Regional Trade Agreements.[2] In the environmental field alone, over 1190 multilateral agreements are presently listed in the International Environmental Agreements database.[3] In other words: a significant amount of multilateral, bilateral and regional agreements are being concluded every year. New agreements are continuously being stacked on top of older ones, a phenomenon that has been denounced as "treaty congestion."[4] In these circumstances, institutional overlaps are almost unavoidable. Institutions can undermine each other, such as when ozone-depleting gases banned under the Montreal Protocol were replaced with gases that do not affect the ozone layer but contribute to global warming.[5] Institutions may also synergize, for instance when instruments for the conservation of biological diversity simultaneously enhance forest carbon sinks. Old institutions can limit the discretion governments enjoy in creating new institutions, such as when the drafters of environmental treaties intended to regulate different types of cross-border flows must avoid conflicts with existing rules under the WTO.[6]

Then there is the issue of inter-institutional order. Ocean governance, for example, proceeds through regional seas agreements covering, for instance, the dumping of toxic waste, pollution from land-based sources or cooperation in cases of oil spills; Regional Fisheries Management Organizations, with authority over territorial waters, Exclusive Economic Zones and parts of the high seas; multilateral institutions with almost global coverage such as the London Convention on ocean dumping and its protocol and the International Convention for the Prevention of Pollution from Ships, consisting of six different annexes covering oil pollution, noxious liquid substances, harmful substances carried in packaged form, sewage, garbage and exhaust from engines and cooling systems; and a host of relevant policy processes under the Food and Agriculture Organization, the International Maritime Organization and the Convention on Biological Diversity, to name but a few. Persistent Organic Pollutants, a class of highly toxic chemicals, fall under different regional agreements covering emissions from stationary sources (a 1998 protocol to the Convention on Long-range Transboundary Air

Pollution) or hazardous waste (the Bamako Convention for Africa or the Waigani Convention for several Pacific Islands) or pesticides and industrial chemicals (the Rotterdam Convention). Climate governance is addressed under the United Nations Framework Convention on Climate Change, its Kyoto Protocol and Paris Agreement; the World Bank; the International Civil Aviation Organization; the International Maritime Organization; and many others. Intellectual property rights fall within the ambit of the WTO's TRIPS Agreement; the World Intellectual Property Organization and the treaties it administers; the World Health Organization; various Free Trade Agreements and so on.

Those multi-institutional settings sometimes evolve in very different ways. Ocean governance, today, lacks a centralized institution with large membership and a comprehensive mandate for addressing the various causes and consequences of marine environmental degradation. Instead, sectoral and regional approaches dominate. Persistent Organic Pollutants were regulated in the same manner until the 2001 Stockholm Convention created a comprehensive regime addressing the substances' production, use, trade, emission and disposal. Climate governance has mushroomed from merely a handful of relevant institutions in the early 1990s to dozens, if not hundreds, of institutions covering everything from emissions trading and technology transfer over adaptation and finance up to emissions from forestry, scientific cooperation and techniques for sequestering carbon dioxide in the oceans.[7] Intellectual property rights was an obscure issue in the early 1980s, became the preeminent domain of the World Trade Organization in the early 1990s and spilled over into institutions addressing development, agriculture and health while simultaneously decentralizing into various trade agreements.

How we can explain changes in multi-institutional governance architectures, and why do different architectures change in different ways? This book is an attempt at a partial answer. In a first step, I identify a particular type of institutional change in arrays of "partially overlapping and nonhierarchical institutions governing a particular issue-area"[8] or "regime complexes." I refer to this type of change as institutional layering, which I contrast with that of club cooperation. In a second step, I develop a novel theoretical account revolving around the concept of situation structure. This approach holds the promise of explaining both types of institutional change in regime complexes, that is, club cooperation *and* institutional layering. While I return to the broader implications of the situation-structural approach in the conclusions of this book, empirically I limit myself to providing a revisionist explanation of what is arguably the best-known case in the literature, the regime complex for genetic resources. Third, in doing so, I test the situation-structural hypothesis against two competing explanations: regime shifting and interplay management.

Two types of institutional change

The diversity of institutions and multi-institutional settings is matched by the diversity of concepts employed in the academic literature, of which "regime complexes" is but one. Some authors prefer to speak more broadly of (multi-institutional) governance architectures, between the elemental institutions of which can exist different degrees of conflict, cooperation or synergy.[9] Others distinguish between "fully integrated institutions that impose regulation through comprehensive, hierarchical rules" and "highly fragmented collections of institutions with no identifiable core and weak or nonexistent linkages between regime elements" as the extreme poles of a spectrum along which we can place different types of institutional settings. Here, regime complexes, or "non-hierarchical but loosely coupled systems of institutions" are towards the middle of the spectrum.[10] A different way to define a regime complex has been "a network of three or more international regimes that relate to a common subject matter; exhibit overlapping membership; and generate substantive, normative, or operative interactions recognized as potentially problematic whether or not they are managed effectively."[11] Others prefer to use "institutional complexes" to refer to roughly the same phenomenon.[12] Still others use the terms "complexity" and "fragmentation" largely synonymously.[13]

Club cooperation

This at times confusing conceptual diversity notwithstanding, several authors have raised the questions of what causes regime complexes to emerge in the first place, and how they change. The two questions are strongly related. The transition from, say, the governance of a particular issue area through a single international institution to an arrangement where multiple and overlapping institutions exert regulatory authority is a question of both emerging complexity and institutional change. The primary explanation found in the literature is what I refer to here as "club cooperation": some members of a given institution consider existing rules as out of step with their interests yet, being unable to reform the institution (due to the presence of veto players, for instance), they choose to create alternative settings for cooperation among like-minded actors. By reducing the variance of interests among participants, club settings allow for the creation of stricter (or simply different) rules than would be possible in a setting where interests are more heterogeneous.

While in the past many scholars have warned of the growing fragmentation of international institutions and international law, clubs allow for a pick-and-choose approach to international cooperation where comprehensive deals with universal participation are impossible. Take the global governance of chemicals and waste. In addition to countless regional agreements, three broader conventions have been put in place since the late 1980s: the Basel Convention on the Control of Transboundary Movements of Hazardous Wastes and Their Disposal; the Rotterdam Convention on the Prior Informed Consent Procedure for Certain Hazardous Chemicals and Pesticides in International Trade and the Stockholm Convention on Persistent Organic Pollutants. Turkey is not a party to the Rotterdam Convention but has joined Stockholm and Basel. Italy has joined Rotterdam and Basel but not Stockholm. The US has joined none but has nevertheless decided to ratify the more recent Minamata Convention on Mercury, whose entry into force is pending. The example highlights the impossibility of aggregating the preferences of all actors within a single, comprehensive institution. The alternative to club cooperation, then, would be the absence of any kind of multilateral regulation whatsoever.

Club cooperation is thus driven by interest differentials. One study links the emergence of the regime complex for energy to fluctuations in the price of oil on the world market, leading to either importers or exporters of oil becoming dissatisfied with existing institutional arrangements at different points in time, thus choosing to create alternatives.[14] "Complexity" results from the fact that the institutions created in such a manner all deal with different issues within the wider issue area of energy. In the case of hazardous waste, several regional agreements have emerged since the 1990s as a reaction to the perceived insufficiency of the multilateral Basel Convention, leading to an overlapping system of institutions taking different regulatory approaches to the transport of waste across international borders.[15] Or take Genetically Modified Organisms, where importing states considered the available rules for risk assessment and import restrictions under the WTO's Agreement on Sanitary and Phytosanitary Measures as insufficient, opting for a more comprehensive approach under the Cartagena Biosafety Protocol to the Convention on Biological Diversity, to which large exporters of transgenic food and feed never acceded.[16] In all those cases, complexity arises because the new institution regulates similar matters as the old institution.[17]

Institutional layering

Some actors' inability to have their way in a given institution causes them to pursue alternatives. Yet this is not the whole story. What we see in many cases is that actors who are dissatisfied with the status quo (or "reformist actors choose *not* to create *à la carte* rules in exclusive clubs. Instead, there are usually protracted and long-winded attempts at modifying or adding to the rules of existing institutions encompassing both reformist and status quo-oriented actors. Here, given interest differentials, reformist actors must scale back their demands for institutional change to obtain the approval of status quo-oriented ones; offer issue linkages in order to seduce the latter into agreement; or both.

This is the type of institutional change I call "institutional layering." The purpose of institutional layering differs from that of club cooperation. The latter aims at closely aligning the rules agreed on in a club setting with the interests of the club members. This is done by limiting the variance of interests among participants. The former seeks to change the rules that apply not only to *reformist* but to *all* actors through the "active sponsorship of amendments, additions, or revisions to an existing set of institutions."[18] On one hand, this may require concessions in order to avoid the participation-limiting effect of deeper cooperation.[19] On the other, it may require issue linkages, that is, the connecting of the adoption of certain rules to the priority issues of status quo-oriented actors in other areas.[20]

The concept of institutional layering has not been systematically developed for the study of institutional change in regime complexes.[21] Its origins are in comparative politics where it has been developed to address both questions of institutional change and continuity. The concept links to wider debates on path dependence in historical institutionalism. While subject to different uses,[22] layering is a strategy adopted by reformist actors when they are unable to displace or transform existing institutions against the opposition of status quo-oriented actors who have "access to institutional or extra-institutional means of blocking change."[23] Reformist actors thus create novel institutional layers on top of an existing regulatory framework in order to specify, clarify, elaborate, adapt or implement existing norms, principles, rules or decision-making procedures.[24] This process is not merely an expression of the increasing precision of international law.[25] Instead, institutional layering is deliberately aimed at bringing international institutions in line with reformist actors' interests.

Club cooperation and institutional layering link to regime complexity in different ways. Clubs are genuinely new institutions that overlap with the original institution with which reformist actors were dissatisfied yet were unable to change. For instance, African countries unable to obtain a comprehensive ban of North-South trade in hazardous waste under the Basel Convention thus chose to create the regional Bamako Convention, which categorically prohibits imports into the treaty area. The unwillingness of certain members of the International Monetary Fund to agree to a reform of quotas and voting power led to several emerging economies creating new development banks, leading to novel overlaps within the issue area of multilateral development finance.

Institutional layering, conversely, leads to overlaps when the novel sets of rules grafted onto existing institutions broadens their functional scope into new areas. A recent amendment to the London Protocol on ocean dumping and a decision by the Conference of the Parties to the Convention on Biological Diversity led both institutions to overlap in regards to ocean iron fertilization for climate geoengineering.[26] The same goes for pharmaceutical patents, originally the exclusive domain of the WTO's TRIPS Agreement. Here, a regime complex emerged when "[a] coalition of developing countries, supported by public health NGOs, struggled successfully to put drug-related intellectual property rights on the WHO agenda, and thereby deliberately created functional overlap with the WTO."[27] Table 2.1 summarizes the differences between the two types of institutional change.

The link to path dependence

Institutional layering is a way of working around the constraints resulting from actors that are crucial for international cooperation yet tend to prefer the status quo. Club cooperation is essentially the opposite: limiting participation to likeminded actors allows deeper cooperation and/or eliminates the necessity of incorporating issue linkages for ensuring the consent of status quo-oriented actors. Each results in a different type of institutional change. Club cooperation can lead to the rapid proliferation of multilateral, minilateral, regional or sectoral regimes. Institutional layering is a more inert type of change, consisting of slow-moving,

Table 2.1 Types of institutional change

	Objective	*Mechanism*	*Impact on other institutions*
Club cooperation	New rules for reformist actors	Limit interest differentials among participating actors	Clubs generate overlaps
Institutional layering	New rules for all actors	Induce broad cooperation through concessions and issue linkages	Layers generate overlaps

gradual processes that are politically more contentious due to the necessity of finding common ground among actors who may have very different ideas as to the direction in which to develop existing institutions.

Club cooperation and institutional layering thus produce different types of path dependence.[28] The former leads to a punctuated equilibrium pattern or critical junctures: in response to an external shock, a governance architecture rapidly shifts to a new equilibrium, over time resembling a pattern of long periods of inertia interspersed with sudden bursts of change. Institutional layering leads to incremental and endogenous change instead: rather than being driven by external shocks, change results from reformist actors' slowly building upon existing institutional frameworks. Incremental change thus results from the constraints the underlying situation structure imposes on reformist actors.

Both types of change are associated with different types of path dependence—a concept that is subject to wildly differing interpretations and has made little inroads in international relations;[29] yet the language of path dependence has been widely adopted in the literature on regime complexity. For instance, structural and institutional constraints may prevent far-reaching changes and "privilege path-dependent future developments following existing patterns."[30] Regime complexes are "broad structures that do not easily move off current trajectories";[31] they possess "considerable stability or equilibrium" due to "interest- and power-based path dependence";[32] or path dependence results as "the cost of changing strategies increases as time passes."[33] All those authors equate path dependence with slow-moving change. The growing cost of veering off a chosen path is especially reminiscent of path dependence as a result of positive feedback: choices are self-reinforcing because the longer they are being pursued, the higher the costs of switching to a different path become.[34]

But this is only half the story. In fact, others have explicitly opted for a punctuated-equilibrium model in which change is not incremental but abrupt and a reaction to exogenous shocks.[35] Punctuated equilibrium is a form of path dependence as well. Yet rather than institutions' slowly converging on an attractor point, change proceeds through branching paths: exogenous shocks require actors to make institutional choices that remain in place until they reach the next fork in the road. While both types of institutional change are considered forms of path dependence, they are diametrically opposed concepts. Why did a broad coalition of developing countries push for a TRIPS amendment on parallel imports for patented pharmaceutical products rather than switch to an alternative and exclusive venue where status quo-oriented industrialized countries could not slow down the process and limit the range of potential institutional outcomes? Why did developing countries push for a development agenda in regards to intellectual property rights in the World Intellectual Property Organization where industrialized countries' opposition is strong, and why did they not choose cooperation among the G77/China instead? Why do inter- and transnational institutions for climate adaptation and finance, emissions trading, technology transfer and carbon sinks spread like wildfire, yet there is only a single international institution containing legally binding targets for greenhouse gas emissions? Why did the states that have the largest stakes in emissions reductions choose to go along with a multilateral UN process that has arguably delivered relatively little over the course of more than two decades, and why have they not chosen to exclude the laggards and cooperate through clubs instead? Why has the EU consistently offered side payments, opt-out clauses or financial assistance to regulatory laggards in order to achieve broad, multilateral cooperation on chemical pollution, and why has it not chosen instead to simply go along with like-minded countries only? The argument I develop in this book is a first step towards explaining those differences.

Explaining institutional layering

The theoretical approach I develop here is applicable to both change through club cooperation and institutional layering, yet, while I address questions of generalization in Chapter 9, this book focuses on explaining the latter. What I will be doing in subsequent chapters is thus not a comprehensive test but is rather intended as an initial exploration of the theory's explanatory power for one specific type of institutional change. The literature on regime complexity does not use the term institutional layering, but it includes two theoretical approaches that are similar in nature and could equally provide an explanation for the creation of novel sets of rules on top of existing institutional frameworks. Before I elaborate on my own approach, I will consider those two in turn.

Regime shifting

The early literature emphasized conflict as a major driver of institutional change in regime complexes. "Regime shifting" arises when states (and non-state actors) "relocate rulemaking processes international venues where whose mandates and priorities favor their concerns and interests,"[36] yet the concept may also entail the creation of strategic inconsistency whereby states attempt to "force change by explicitly crafting rules in one elemental regime that are incompatible with those in another."[37] Regime shifting thus has two different sides: one in which actors aim to craft novel rules in favorable venues, another where actors attempt to undermine the operation of target institutions. The former implies that regime complexes change as actors switch to their preferred institutions; the latter that they result from "contestations between regime actors who are using linkages between issue-areas as legitimating frames through which to change or subvert" a target institution.[38] Strategic inconsistency can play out both at the level of rules, that is, "specific prescriptions or proscriptions for action" and at the level of *norms*, "standards of behavior defined in terms of rights and obligations."[39] While it happens rarely, if ever, that states adopt rules that blatantly contradict existing parts of international law, strategic inconsistency can serve to challenge and de-legitimize the status quo or to strengthen institutions that are threatened by the preponderance of more-powerful ones.[40]

The conceptual boundaries of regime shifting, forum shopping and strategic inconsistency are not particularly clear as different authors use quite different definitions and approaches.[41] As such, the relocation of policy processes from one forum to another need not lead to inconsistencies, even "strategic" ones. Yet inconsistencies can easily arise when a given issue is simultaneously subjected to different regulatory approaches or "frames" under multiple institutions.[42] Regulating genetically modified food from a perspective of trade liberalization is one thing. Regulating it under the vantage point of protecting the environment from genetic pollution and human health from presently unknown adverse effects is another, and actors choosing to pursue the latter approach are obviously dissatisfied with the former one and would very much welcome it if free trade in genetically modified food were, to some extent, rolled back. Even without leading to manifest legal conflict, regime shifting can undercut established institutions in several ways. It can spur broader policy processes that lead to the adoption of specific sets of rules further down the road. By shifting to the World Health Organization during the 1990s, developing countries and civil society organizations challenged international patent law and the restrictions on access to essential medicines it engenders; subsequently, this led to more-specific attempts at balancing concerns over intellectual property with concerns over public health.[43]

Regime shifting can also challenge existing institutions by creating inconsistencies not at the level of rules, but at the level of their implementation. This phenomenon can be observed in refugee politics. Since the 1990s, the Refugee Convention has faced increasing competition from international regimes in the areas of international migration and internal displacement. At the level of rules, there are no conflicts here: the three issue areas are legally unrelated. Yet international obligations for states to grant protection to refugees are only triggered once they reach their respective territories. By creating regimes in the unrelated areas of international migration and internal displacement, some Northern states have attempted to prevent, or at least complicate, the arrival of refugees to their territories in order to keep those obligations from becoming active.[44]

I thus broadly understand regime shifting as the strategy of bringing an issue under a new regulatory approach, leading to greater or smaller inconsistencies at either the level of rules or norms or their implementation. Institutional change is thus largely driven by political conflict. Below I will elaborate on important similarities to my own approach. At the same time, the concept of regime shifting is the exact counterpart of a more-recent and increasingly dominant approach, namely interplay management.

Interplay management

Empirically, the extent to which states generate legal inconsistencies is probably overrated. The recent academic literature has been developing in the opposite direction. While states may have interests in weakening or undermining certain international regimes, they simultaneously have interests in ensuring that the international institutional order does not become dysfunctional.[45] Active sponsorship of inconsistent rules and norms may also reverberate across the wider institutional landscape, damaging an actor's reputation and leading others to retaliate in kind. Instead, states may engage in interplay management: "deliberate efforts [...] to address and improve institutional interaction and its effects."[46] By enhancing the way in which two or more international regimes are coordinated with each other, states avoid negative effects from one regime to the other.[47]

The assumption behind interplay management is that states not only have interests in changing individual institutions with which they might be dissatisfied, but also look at the broader picture: the potential effects their actions in a given setting may have elsewhere.[48] Subverting a particular institution might lead to gains in a specific policy field yet may cause repercussions in others. Short-sighted behavior thus does not necessarily pay off. Interplay management thus does not presuppose normative commitments to effective multilateralism or the order of global governance as such. Rather, rational actors will increase their gains from cooperation if they do not jeopardize the functionality of other institutions through regime shifting.

The interaction between the ozone- and climate regimes is a prime instance of interplay management. When ozone-depleting industrial gases were phased out under the Montreal Protocol and replaced with gases that are benign to the ozone layer yet have adverse effects on the climate system, governments extended the Protocol's regulatory scope to reduce negative impacts in the area of climate governance.[49] The MARPOL and Basel Conventions provide another instance of interplay management. The former regulates vessel-source pollution, the later the generation and transboundary movements of hazardous wastes with the exception of wastes resulting from the normal operations of ships. The 2006 *Probo Koala* incident, when an industrial process resulting in the generation of hazardous wastes was outsourced to a ship in order to bypass the Basel regime, led governments to embark on an extended process for identifying and closing any regulatory gaps existing between the two agreements.

Interplay management thus seeks to ensure the smooth functioning of international institutions, including the creation of novel sets of rules. States have a "shared interest in avoiding latent and especially manifest normative conflict and implementing pre-existing commitments."[50] We should thus expect them to clarify the division of labor between various regimes by delineating spheres of regulatory authority, and we should expect them to avoid inconsistencies at the level of rules (outputs) or the behavioral changes those rules imply (outcomes).[51] Change, in the form of institutional layering, happens when interplay-management decisions lead to marginal adjustments of existing rules and amendments or the creation of implementing agreements.

Situation structure

The approach I develop and test here is somewhat similar to the idea of regime shifting yet is more-strongly geared towards actors seeking to maximize their gains from cooperation under constraints. The term "situation structure" broadly captures "the strategic nature of the situations in which states make choices about cooperation."[52] First, this is a matter of the specific patterns of *interdependence* between states. In transboundary pollution, the downstream state is dependent on the upstream state to cease the polluting activity, yet the same does not hold true in reverse.[53] In illegal trade in hazardous wastes, net importers require international regulation and oversight more than net exporters. In international lending, poor states need wealthy states to provide financial resources more than the other way round. And in non-proliferation, states that do not possess military or civilian nuclear programs are more dependent on cooperation than those that do.

The first two are examples of negative externalities: by pursuing an activity beneficial to itself, an actor imposes costs on others.[54] The latter two examples are different types of public goods, the supply of which is beneficial both to the supplier and to others. There is another important aspect here pertaining to excludability. A downstream state cannot effectively prevent transboundary pollution without the cooperation of the upstream state; yet in the case of illegal waste, better import controls might get the job done even without the state of export taking any measures. States can effectively be excluded from international lending yet not from the benefits resulting from non-proliferation policies.

Patterns of interdependence are a major source of influence in international politics.[55] Yet, in addition to interdependence, my concept of situation structure also encompasses *interests*. More specifically, interests are partially endogenous to situation structure. This is where we get to basic problems of collective action: a state generating negative externalities usually benefits from doing so and thus prefers the status quo over regulatory cooperation. Poor states would prefer wealthy states to provide more financial assistance and better conditions, yet wealthy states have to foot the bill. This does not necessarily imply that states capable of supplying relatively large amounts of goods are oriented towards the status quo. A large body of literature shows that states may supply non-rivalrous goods to the wider international community unilaterally or minilaterally[56] or leverage their capacities to provide structural leadership in a multilateral setting.[57] Under the approach adopted here, however, this requires that the costs of doing so are below the benefits they themselves receive by doing so.

The link to institutional change is this: situation structure determines which actors are set to gain from institutional changes and which are bound to incur costs. This shapes the distribution of interests, with some actors being *demandeurs* of institutional change (reformist), others preferring things to stay the way they are (status quo-oriented) and still others somewhere in the middle. Situation structure then determines the *type* of institutional change. Reformist actors cooperate through clubs where goods are excludable (to prevent status quo-oriented actors from free riding on their contributions) or negative externalities avoidable (to create deeper rules than would be possible if status quo-oriented actors were to participate). Reformist actors attempt to cooperate through multilateral settings where goods are non-excludable (to have status quo-oriented actors provide more than they would do otherwise) or negative externalities unavoidable (to have them reduce more than they would do otherwise). As noted above, the former type of institutional change allows deep cooperation by limiting the range of interest differentials whereas the latter type requires shooting for shallower cooperation and/or incorporating issue linkages to convince status quo-oriented actors to join in.

While this book largely limits itself to explaining one particular type of institutional change, namely layering, the implications of the situation-structural approach are potentially broader. The approach can explain, for instance, why dissatisfaction with the Basel regime on transboundary waste led South-American, African and Pacific states to create regional waste treaties—as negative externalities from Northern waste exports are avoidable through tighter import controls, multilateral cooperation is not necessary. The approach also explains why, conversely, states looking for comprehensive cooperation on Persistent Organic Pollutants chose multilateral cooperation under the Stockholm Convention—with negative externalities from long-range, transboundary pollution being unavoidable, the chief generators (here: particularly countries from the Global South) had to be convinced to join a multilateral cooperative arrangement—requiring both linkage (financial and technical assistance) and the trading-off of depth for participation (the flexibilities the Stockholm Convention allows for).[58] I return to those wider questions in Chapter 9.

So what are the differences to the regime-shifting approach? First, situation structure implies that actors may regime-shift if they expect this to increase their gains from cooperation. Yet when are such shifts successful, when are they partially successful and when do they fail? On one hand, this is a matter of bargaining leverage, which I derive here from asymmetrical interdependence. On the other, it is a question of the interests of those status quo-oriented actors to be included. Where sufficient issue linkages are available, they may be persuaded to accept bringing a given within the functional scope of another institution, thus leading to the emergence of overlaps. In other cases, reformist actors may be required to scale back their ambitions. Then regime shifts may be partial and merely result in, say, the adoption of some rather unspecific rules within another institution with unclear relevance for the issue at hand. Chapter 8 delivers an example of this, where attempts to bring viral genetic resources within the ambit of state sovereignty were only partially successful. The case studies also show that there are important overlaps between regime shifting and situation structure that, at times, make it difficult to adjudicate between the two.

Final remarks

No theory can explain everything. As the empirical chapters show, this includes the situation-structural approach. First, situation structure can only partially account for the variations in state interests. Other factors, including normative ones, may be relevant as well and can go a long way towards explaining bargaining behavior, institutional preferences and outcomes. Second, actors' relative positions do not *determine* the outcomes as "bargaining is usually a means of translating potential into effects, and a lot is often lost in the translation."[59] The point of the exercise is to maximize inferential leverage: explaining a broad variety of empirical phenomena with as small a set of theoretical assumptions as possible.[60] In this sense, the approach is deliberately primitive. Second, as noted above, this book is not a comprehensive theory of the situation-structural account but limits itself to a preliminary exploration in regards to explaining one particular type of institutional change, namely, institutional layering. Arguably, this is where the approach's real challenge lies as it faces powerful competition from alternative explanations. As I discuss in Chapter 9, its application to club cooperation is significantly easier.

Methodology

In the following chapters, I deal with what is probably the archetypal instance of a regime complex, the case of genetic resources, which has been the subject of several case studies. Interestingly, institutional changes in this regime complex have been explained in terms of both regime shifting and interplay management. A seminal paper considers the design of two key institutions, the CBD and the FAO Seed Treaty, to be strongly shaped by the attempt to create counter-regime norms against the privatization and commercialization of plant genetic resources.[61] Another study argues that the complex evolves from fragmentation to competition, specialization and integration due to systemic pressures on governments to ensure policy coherence across its elemental institutions. This, in turn, feeds back to the level of the complex itself.[62] Finally, the Nagoya Protocol, a key institution within the complex, has been described as a prime example of successful interplay management.[63]

The genetic resources regime complex consists of overlapping institutions from areas such as intellectual property rights, biodiversity, agriculture, health and ocean governance. At its core, it revolves around the questions of who is granted access to different types of genetic resources, under which conditions, to what extent do the benefits resulting from the (biotechnological) utilization of such resources need to be shared, and with whom? Access and benefit-sharing (ABS) is governed by multiple, overlapping institutions that are sometimes at odds with each other yet, at other times, have entered into intricate divisions of labor.

In Chapters 5 to 8, I focus on four different processes of regime formation. Chapter 5 deals with the regime for Plant Genetic Resources for Food and Agriculture, which, with the 2001 FAO Seed Treaty, resulted in an ABS mechanism that grants access to a transnational network of seed banks for plant breeding while requiring that the benefits arising out of the utilization of the materials provided are shared multilaterally. Chapter 6 covers the negotiations on a dedicated regime for combating biopiracy, an ambiguous and problematic term to which I shall return in Chapter 4. This regime essentially deals with monitoring and compliance: it seeks to ensure that the benefits from the utilization of genetic resources are shared fairly and equitably, as set out in the relevant international treaties. Chapter 7 deals with pandemic influenza viruses and the formation of an ABS regime under the WHO. Here, the multilateral sharing of candidate viruses for rapid vaccine production during a pandemic emergency is linked to the multilateral sharing of resulting vaccines and other pharmaceutical products. Chapter 8, finally, deals with a moving target: the ongoing negotiations on an ABS regime for marine genetic resources in Areas Beyond National Jurisdiction. This process is linked to much broader discussions on how to protect the marine environment of the high seas.

This case selection is largely exhaustive. The first three cases cover all international ABS regimes currently in existence, and negotiations on an ABS regime for marine genetic resources are currently progressing at a rapid pace. There are two cases I do not cover. The first is the negotiation process for an ABS regime for genetic resources and traditional knowledge that has been ongoing for more than 15 years with remarkably little to show. This case also overlaps in significant ways with the debate on an international regime for combating biopiracy, which is why I cover it in Chapter 6. The second potential case is the debate on ABS from animal genetic resources, which presently exists merely in the form of thought experiments and informal discussions, rather than actual negotiations.

In each case covered in this text, I assess the situation structure in four different ways. First, intellectual property claims by geographical origin represent countries' relative capacities to utilize genetic resources and generate the benefits that may (or may not) be shared fairly and equitably after the fact. They also determine countries' relative stakes in ensuring access to genetic raw materials for generating commercial value and, depending on the circumstances, may cause transnational negative externalities due to third parties' being excluded from using, selling, producing or exporting the protected inventions and plant varieties. Second, the global distribution of the different types of genetic resources represent the relative importance (or lack thereof) of countries as providers of the biological raw materials for the subsequent generation of benefits. Third, I address the feasibility of access regulation: can the utilization of different types of genetic resources effectively be prevented, thus giving providers leverage? Fourth, I cover the relevant technological infrastructure necessary for generating benefits from utilization. This includes seed banks in the case of plant genetic resources for food and agriculture, surveillance capacities for viral genetic resources and technologies for accessing marine genetic resources on the high seas or the ocean floor. From the overall situation structure, I derive different groupings of actors depending on their respective roles in providing and utilizing genetic resources and deduce their respective interests, which I assume to be homogeneous within the respective groups. Again, this assumption is extremely simplifying. As the cases will show, there are important variations between actors that cannot be accounted for by situation structure. The reason for this and other simplifying assumptions is, to stress this once more, the development of a parsimonious theoretical approach capable of explaining not everything about every case, but a large number of phenomena relative to its sparse set of assumptions.

The empirical core of the four case studies is the analysis of the relevant negotiation processes. I base my empirical analysis, first, on primary documents from the various negotiation processes. This entails parties' submissions, chairs' summaries and negotiated text. One technique I use is to look at how draft treaty texts evolve over time. As language that has not been agreed upon is usually marked, this allows me to identify the precise timing at which consensus emerged. Second, I include various negotiation reports from third-party observers. Those are chiefly the Earth Negotiation Bulletins published by the International Institute for Sustainable Development, as well as reports by Intellectual Property Watch. I also include a limited number of US embassy cables that give insight into the internal deliberations and assessments in the US government. Where appropriate, I also draw on the secondary literature.

For clarity's sake, I chose to use a structured and roughly chronological narrative that focuses on singular issues and package deals. Each case study concludes by testing the explanatory power of regime shifting, interplay management and situation structure for both institutional outcomes and the processes leading up to those outcomes.

Congruence procedures, process tracing, observable implications

I understand "institutional outcomes" in terms of the formal rules pertaining to access, benefit-sharing, compliance and scope, as well as in the manner in which different norms related to ownership of and control over genetic resources are codified within the respective regimes. I understand "negotiation processes" as bargaining over such rules and norms. In the case studies, first, I use congruence procedures to match institutional outcomes to theoretical expectations. For regime shifting, I ask how greatly institutional outcomes undermine other regimes and the division of labor with them at the level of outputs and outcomes. For interplay management, I look into how far outcomes resolve problematic interactions and create clear divisions of labor. For situation structure, I assess how far the outcomes are reflective of actors' positions in the situation structure. By grouping together countries (and regional organizations) based on the goods they are capable of supplying (genetic raw materials and shared benefits arising out of their utilization) and the negative externalities they generate (through intellectual property claims), I deduce both their interests and relative dependence. Outcomes should closely align with the interests of the least-dependent groups of actors. While there is no direct way to test how interests weighted by interdependence translate into specific institutional outcomes, congruence procedures can approximate whether or not outcomes are largely reflective of the interests of those actors in the most-privileged positions in the situation structure.

Second, and where available, I use "diagnostic evidence" to examine the "observable implications of [the] hypothesized causal mechanisms [...] to test whether a theory on these mechanisms explains the case."[64] The data points on which I focus are the linkages actors propose and agree to; shifts in their bargaining positions; and the timing of such linkages and shifts. A linkage makes agreement on one issue contingent on the agreement on another. It thus involves mutual concessions. Linkages also entail trade-offs where agreement on one issue is made contingent on *foregoing* agreement on another. Shifts in negotiating positions are clearly identifiable instances where an actor changes its demands in regards to the rules and norms to be included in the respective regime. Such shifts can form part of a linkage, but they can also entail the unilateral surrender of parts of an actor's negotiating position, for instance in order not to clog up a negotiation process with unrealistic demands or facilitating broader package deals in the future.

Regime shifting, interplay management and situation structure entail different observable implications: things we should be able to see if each of them was true:

- Regime shifting implies that reformist actors attempt to bring new (and, for them, more advantageous) regulatory approaches to bear on a given set of issues or that they attempt to craft rules that undermine a target regime they consider disadvantageous to themselves. Both entail the creation of rules that weaken or even displace existing regulatory approaches. Institutional design choices will thus encompass the blurring of the divisions of labor between institutions. This entails expanding the functional scope of an institution into areas in which other institutions already exert authority; creating norms that (partially) conflict with those under a target institution or creating rules that lead to ambiguous rights and obligations in regards to a particular issue. For instance, the concept of Farmers' Rights in the context of the Seed Treaty had originally been an attempt to weaken the regime for plant variety protection in regards to farm-saved seeds. The unilateral concession of much of its normative content is a strong data point against regime shifting in the case of the Seed Treaty. Conversely, the unwillingness of developing countries to compromise on the principle of state sovereignty over genetic resources in the negotiations on the Pandemic Influenza Preparedness Framework strengthens the hypothesis—the eventual institutional outcomes led to a situation in which the obligation of WHO member states to unconditionally share viral materials, and the right of vaccine manufacturers to unconditionally use them, has been put in question.
- Diametrically opposed to regime shifting, interplay management means that actors attempt to implement their pre-existing commitments under different institutions in a complementary manner and thus prioritize the elimination of inconsistencies between regimes and the establishment of clear divisions of labor. This also entails the drafting of unambiguous rules, as ambiguities can lead to inadvertent tensions between different regimes if legal interpretations differ among parties. US opposition to mandatory benefit-sharing

from marine genetic resources in the high seas can be understood as an attempt to prevent inconsistencies between ABS and the freedom of the high seas under part VII of the United Nations Convention on the Law of the Sea—an observation that, however, is equivalent with what situation structure predicts. I return to this problem below. Conversely, the change in the EU's position, from alleged incompatibility of an ABS regime with the United Nations Convention on the Law of the Sea to the acceptance of such a regime, given that it be treated as part of a larger package, implies that the underlying motivation was not concern over inconsistency but rather a wish to create leverage for obtaining the support of the G77/China in regards to the protection of marine biodiversity.

- Situation structure holds that actors attempt to maximize their expected gains from international ABS regimes: user countries prefer uncomplicated access to genetic resources without strings attached; provider countries seek to obtain maximal bilateral backflows; non-users/non-providers prefer to limit negative externalities from intellectual property rights or, where those do not exist, arrangements for multilateral benefit-sharing. On one hand, situation structure implies that actors disregard problems of inconsistency and unclear divisions of labor insofar that this increases their expected gains. On the other, actors will concede or trade off their demands for institutional design elements that could lead to inconsistencies or unclear divisions of labor if doing so increases the gains from the resulting regime. In the negotiations on the Seed Treaty, for instance, developing countries' consent to the patentability of certain plant genetic resources for food and agriculture in exchange for the sharing of resulting benefits is a case in point. Conversely, situation structure faces difficulties in explaining the EU's shift towards a proactive stance on ABS from marine genetic resources due to their commercial relevance for the fledgling European blue biotech industry.

Timing, finally, allows for additional inferential leverage in cases where linkages and shifting negotiating positions form part of structured negotiation processes with a foreseeable end date. Data points against a particular hypothesis weigh more heavily the earlier they occur. The reason is that we should expect actors to maintain the priority issues in their negotiating position as far into the process as possible in order to hold out for a possible deal at a later point. Those parts of their negotiating positions that are traded off or conceded early on are, accordingly, low-priority issues.

Counterfactuals

This text focuses on the narrower question of how to explain institutional layering in the genetic resources regime complex, yet the broader question is: why did institutional change proceed through layering and not through club cooperation? For answering the latter, I use counterfactual analysis:[65] what would have happened if the different groups of actors had chosen club cooperation instead of institutional layering, thus obviating the need for scaling back ambition and using linkages to get others on board in a multilateral setting? The use of counterfactuals thus allows for a controlled thought experiment: would actors have been better off if they had chosen clubs instead of going down the multilateral route, or not? The exercise thus allows for the comparison of factual with counterfactual cases.[66] If the situation-structural explanation holds true, actors should prefer the factual world (of institutional layering) to the counterfactual one (of club cooperation).

For constructing those counterfactuals, I first look at the gains from cooperation that club members could have realized among themselves, for instance by making certain types of genetic resources available more easily. Second, I turn to the relationship between club members and non-members. Do non-members impose unavoidable costs on the club, or does their exclusion prevent members from obtaining certain goods, in the form of genetic resources or shared benefits? Third, if some actors would have chosen club cooperation, in all likelihood there would have been an effect on the behavior of *other* actors: a second-order counterfactual.[67] In other words: the net gains to be had from counterfactual club cooperation also depend on the counterfactual behavior of non-members. For this reason, I construct my counterfactuals, first, under the assumption that they would have led to negotiation failure at the multilateral level: a club for virus-sharing would have led to the collapse of WHO negotiations on the Pandemic Influenza Preparedness Framework; a club of large providers of plant genetic resources for food and agriculture would have let the negotiations on the FAO Seed Treaty collapse and so forth. Second, I ask what would have happened if club cooperation had resulted, yet non-members had still chosen to adapt the respective international regime in precisely the same manner as it happened in the factual world.

On a technical note, the application of this method to Chapter 8, where negotiations are ongoing, falls under the label of "scenario analysis" instead. Here, club cooperation is not "contrary to fact" but rather a remaining (yet highly implausible) possibility. While counterfactual analysis deals with things that did not happen in the past, scenario analysis addresses things that could happen in the future.[68] Other than that, the analysis in Chapter 8 takes the same approach as outlined above.

Notes

1 see https://www.icrc.org/applic/ihl/ihl.nsf/vwTreatiesByDate.xsp, January 13, 2016.
2 see http://rtais.wto.org/UI/PublicAllRTAList.aspx, January 13, 2016.
3 See http://iea.uoregon.edu, January 13, 2016.
4 Hicks (1998).
5 Johnson and Urpelainen (2012).
6 Axelrod (2011).
7 Biermann et al. (2009); Keohane and Victor (2011).
8 Raustiala and Victor (2004), 279.
9 Biermann et al. (2009).
10 Keohane and Victor (2011), 8.

11 Orsini et al. (2013), 29.
12 Oberthür and Pozarowska (2013); Gehring and Faude (2014).
13 Gupta et al. (2015).
14 Colgan et al. (2011).
15 Clapp (1994).
16 Gehring and Faude (2014).
17 Urpelainen and van de Graaf (2015).
18 Streeck and Thelen (2005), 24.
19 Gilligan (2004); Bernauer et al. (2013).
20 Haas (1980); Sebenius (1983); Young (1996); Koremenos et al. (2001); Mitchell and Keilbach (2001); Leebron (2002).
21 An earlier version of this argument can be found in Rabitz (2016).
22 See van der Heijden (2011).
23 Mahoney and Thelen (2010), 20.
24 Krasner (1982).
25 Abbott et al. (2000).
26 Ginzky and Frost (2014); Decision XI/20.
27 Gehring and Faude (2014), 489.
28 Mahoney (2000); Beyer (2010).
29 Mahoney (2010); Pierson (2004); Beyer (2010); Hanrieder (2015).
30 Oberthür and Pozarowska (2013), 114.
31 Orsini et al. (2013), 37.
32 Zelli and van Asselt (2013), 9.
33 Struett et al. (2013), 95.
34 Pierson (2004).
35 Colgan et al. (2011).
36 Helfer (2009), 39.
37 Raustiala and Victor (2004), 301–302.
38 Muzaka (2010), 772.
39 Krasner (1982), 186.
40 Helfer (2009), 41.
41 Alter and Meunier (2009); Betts (2009); Helfer (2009); Muzaka (2010); Sell (2010); Rabitz (2016).
42 Muzaka (2010).
43 Muzaka (2010).
44 Betts (2009).
45 Gehring and Faude (2014).
46 Oberthür (2009), 373.
47 Johnson and Urpelainen (2012).
48 Gehring and Faude (2014).
49 Johnson and Urpelainen (2012).
50 Oberthür and Pozarowska (2013), 105.
51 Oberthür (2009).
52 Hasenclever et al. (1996), 187.
53 Mitchell and Keilbach (2001).
54 Cornes and Sandler (1996).
55 Keohane and Nye (2012), 9.
56 Schelling (1973); Snidal (1985); Kaul et al. (2003); Barrett (2007).
57 Young (1991).
58 Lallas (2001).
59 Keohane and Nye (2012), 10.
60 King et al. (1994).
61 Raustiala and Victor (2004), 301–302; see also Helfer (2009).

62 Morin and Orsini (2014).
63 Oberthür and Pozarowska (2013).
64 Bennett and Checkel (2015), 7–8.
65 For counterfactuals in the social sciences, see for instance Fearon (1991); Sylvan and Majeski (1998); King and Zheng (2007); Lebow (2010); Goertz and Mahoney (2012), 115–124. For general background, see Collins et al. (2004).
66 Fearon (1991).
67 Lebow (2010), 51–52.
68 See van Notten et al. (2003).

3 Genetic resources and biotechnology

A confluence of technological, legal and economic changes has driven institutional changes in the global governance of genetic resources. The biotechnological revolution in agriculture, pharmaceuticals and industrial processes, together with the strengthening and international diffusion of intellectual property rights, has transformed entire industrial sectors. Within this wider nexus, the commercial use of biological materials has gained currency: plants could suddenly be engineered for insect resistance or herbicide tolerance; chemical compounds could be produced by modifying the gene expression of micro-organisms; naturally occurring substances could be synthesized and enhanced in the laboratory in order to produce novel drugs; and bacteria could be deployed for industrial fermentation processes and biocatalysis.

This chapter discusses the bearing of technological, legal and economic changes on the utilization of genetic resources. I first address the relevance of such resources for agricultural, pharmaceutical and, to a lesser extent, industrial biotechnology. I then turn to changes in intellectual property law. Those changes have played out on different levels. At the national and regional one, the scope and depth of intellectual property protection has been expanded primarily through case law. At the international level, different treaties have created minimum standards for national intellectual property legislation, leading to the proliferation of stricter protection regimes in most parts of the world. Finally, I discuss economic changes, that is, the evolution of markets for genetic resources and shifts in the structure of relevant industries.

Red, green, white

Defining biotechnology is far from straightforward. It is rather a bundle of technologies and approaches used in a broad range of commercial applications as well as in basic research. Its application is not confined to a particular academic discipline such as biology or genetics, instead overlapping with fields such as informatics, chemistry, agronomy and medicine. Likewise, biotechnology varies in its relevance for different economic sectors and, equally, across countries. For most purposes, it is employed jointly with other technologies, sometimes playing a more central role, sometimes a more peripheral or merely supportive one.

Distinguishing some modern forms of biotechnology, such as recombinant DNA technology, from practices that have been around for centuries or even millennia, is equally challenging: the use of micro-organisms for fermentation has been known to cheese makers and brewers since time immemorial. The Europeans discovered vaccines, the first biopharmaceuticals, in the 18th century; the Chinese did so 800 years before.[1] The crossing of different plant varieties for obtaining useful and stable genetic traits in the progeny has been the basis of agriculture since the Neolithic Revolution some 12.000 years ago. Whereas contemporary technology is infinitely more sophisticated, it is in many ways not qualitatively different from the rather more pedestrian approaches of our ancestors.

Any definition of biotechnology is thus necessarily tentative. The Convention on Biological Diversity defines it as "any technological application that uses biological systems, living organisms, or derivatives thereof, to make or modify products or processes for specific use."[2] The OECD understands it as "the application of science and technology to living organisms as well as parts, products and models thereof, to alter living or non-living materials for the production of knowledge, goods and services"; this may include, for instance, technology for DNA sequencing; genetic modifications for having organisms express useful proteins: cell and tissue cultures for rapidly producing genetically uniform plants or vaccines or for bacteria that have been engineered to express commercially interesting proteins; the application of informatics and engineering to the design of biological systems; vectors for efficiently injecting genetic materials into host cells or even nanobiotechnology for the construction of biological machines.[3]

The same definitional problems apply for the concept at the core of this book: genetic resources. They are legally defined as "any material of plant, animal, microbial or other origin containing functional units of heredity" that is "of actual or potential value."[4] The definition thus includes plant, animal and microbial materials, but also those originating from fungi and the human genome.[5] It also extends to the *products* of such materials, given that they contain "functional units of heredity" and are "of actual or potential value" and even digitalized genetic sequence data that, transcribed into electronic form, retains its hereditary information.[6] This definition is clearly open to charges of ambiguity. Yet, considering that international law moves at a significantly slower pace than technological innovation, the flip side of ambiguity is flexibility, as the definition of "genetic resources" can be adapted to a broad range of novel circumstances its drafters could not have foreseen in the early 1990s.[7]

Modern biotechnology makes use of genetic resources in four different ways. The *in situ* utilization of genetic resources entails the direct extraction of interesting biochemical compounds from, for instance, plants occurring in the wild. *Ex situ* genetic resources have been removed from their natural environment and are being conserved for varying periods of time in gene banks, frequently offering elaborate catalogs for allowing users to search in a targeted way for interesting materials. *In vitro* utilization uses isolated and modified components of an organism for synthesizing useful proteins; finally, *in silico* utilization pertains not to the use of physical materials, but merely their genetic sequence data, for instance to synthesize novel micro-organisms from anorganic components.[8]

The commercial use of genetic resources in biotechnology took off in the 1980s, yet the foundations were laid with the discovery of the DNA helix in 1953 and the invention of genetic engineering during the course of the 1970s. To this day, the basis of genetic engineering is the targeted insertion of DNA molecules into a host organism. Those molecules can either be directly extracted from a source organism or synthesized in the laboratory. Since 1983, the Polymerase Chain Reaction enables sufficient amounts of molecules to be rapidly produced. Source material is inserted into a cloning vector through a combination of "scissors" (restriction enzymes) and "glue" (DNA ligase): the vector DNA is first cut at the point of insertion and subsequently linked together with the desired DNA molecules, thus producing recombinant DNA. The latter is then injected into the host organism where the novel DNA is translated into polypeptide chains, which fold themselves into proteins; those can imbue the host organism with specific traits or be harvested for direct use.

Despite considerable technological advances, recombinant DNA still remains the major contemporary approach to genetic engineering. Before discussing a number of relevant novel technologies below, I first turn to the application of biotechnology in three major sectors: agriculture ("green" biotech), pharmaceuticals ("red") and industrial processes ("white"). I devote less attention to the latter than to the former two due to the lower practical relevance of genetic resources and their international governance to industrial biotechnology.

Green biotechnology

Technology had already transformed agricultural breeding long before genetic engineering came along. In the early 20th century, high-yield hybrid crops created by crossing inbred lines of different species quickly came to dominate corn production in the US and subsequently spread to the rest of the world. Radiation and synthetic chemicals for inducing high levels of genetic mutation for rendering large numbers of useful traits in a short period of time had entered crop breeding in the 1920s. Since the 1930s, artificial chromosome doubling enabled the creation of fertile varieties of (otherwise sterile) crosses.[9] Since the 1970s, recombinant DNA technology allowed breeders to insert specific traits, in a targeted manner, into their crops, such as insect resistance or pesticide tolerance ("input traits") or better nutritional values and higher yields ("output traits"). A transgene would be inserted into a receptor variety, which would subsequently be crossed with elite commercial lines. The Flavr Savr® tomato was the first transgenic crop to hit the supermarket shelves in 1994. Monsanto's transgenic soybeans, engineered for resistance against the company's own herbicide, glyphosate, received market approval around the same time.[10]

In the following years, potatoes, corn and cotton injected with genes coding for resistance against certain pests were widely approved for marketing in the US and elsewhere. Genes of the *Bacillus thuringiensis* (BT) proved particularly popular due to the insecticidal proteins they produce. BT cotton, tobacco, potatoes and other plants could thus be made resistant against common pests such as the spotted and pink bollworms, the corn borer and corn rootworm larvae.

The global acreage of transgenic crops has jumped from zero in the early 1990s to presently around 180 million hectares—roughly 12% of the world's arable land.[11] In some regions, such as Europe, the public reaction to those crops ranges from skepticism to outright hostility.[12] Yet in other parts of the world, such as Argentina, Australia, Brazil, Canada, China, India and the US, transgenic crops have been widely adopted. The US is the largest producer, accounting for more than a third of the global total. The most common types of crops are soybeans and corn in the Americas and cotton in Asia and Africa. As of 2015, one-third of corn and 80% of soybeans grown worldwide are transgenic.[13] Despite attempts to broaden the portfolio of genetic modifications, transgenic crops are overwhelmingly engineered for insect resistance and herbicide tolerance.[14] In 2014, those two traits were found in over 80% of commercial cultivations, while traits for higher yields or improved quality are marginal.[15] The overall track record of transgenic plants is mixed, though. Some studies have identified significantly increased farm incomes, crop yields or reduced insecticide use.[16] Others are less sanguine, with one author noting the "tendency of GM [genetic modification] proponents to refer repeatedly to the 'great promises' GM holds for the future rather than to the real value of past achievements."[17] Evidence suggests that the increased usage of blockbuster herbicides for protecting crops has put evolutionary pressure on weeds and led them to develop resistances as well.[18]

In many ways, agricultural biotechnology does not supplant traditional breeding methods but rather complements them. Even where breeding does not make use of recombinant DNA, biotechnology provides tools for the rapid screening of large numbers of plant materials in order to identify potentially useful traits. Potential candidates can then, subsequently, be bred into stable varieties through conventional breeding methods without injecting foreign DNA molecules. This includes technologies such as high-throughput screening, marker-assisted selection or Targeting Induced Local Lesions in Genomes (TILLING). While DNA recombination is strictly regulated in some jurisdictions, this is not the case for tools that simply allow a more efficient use of traditional crossing and breeding methods.

Red biotechnology

As in agriculture, "red" biotechnology also encompasses both methods for enhancing production processes, as well as the products themselves. Biotechnology is widely being used for the production of small molecule drugs, which have traditionally been produced through chemical engineering. The first such drug to hit the markets was human insulin. Until the 1970s, insulin was either extracted from animals or produced through chemical synthesis, the latter being a cumbersome process that was unable to yield large batches. In the late 1970s, California-based Genentech found a way to have the *E. Coli* bacterion express the insulin protein; by replicating the modified bacteria in microbial cultures, production was scaled up dramatically.[19] Following on the heels of recombinant insulin were human growth hormones. Previously extracted from the brains of deceased donors, having bacteria express hormone proteins significantly expanded their global supply. Cloned interferon, for the treatment of cancer, hepatitis or multiple sclerosis, was first produced through bacterial cultures in 1980 and resulted in a "gold rush in genetic engineering."[20]

Beyond providing more efficient ways of producing small-molecule drugs, biotechnology has also enabled the creation of new biopharmaceuticals, or "biologics." Those are presently the fastest-growing segment in global pharmaceutical markets with a total sales volume of US $86 billion as of 2008. They account for 30% of the pharmaceutical industry's global development pipeline.[21] Biologics are significantly harder to manufacture than small-molecule drugs, and only a handful has entered the marketplace. They are produced from plant or animal materials, micro-organisms or human stem cells and tissues. Monoclonal antibodies are the most prominent biologics, with annual sales in the billions of dollars. Initially developed for killing cancer cells, they are also being applied in the treatment of autoimmune, cardiovascular and infectious diseases. A more recent development in biologics is the advent of "biopharming": the production of recombinant proteins via transgenic plants or animals. In 1990, a globular protein was first grown in genetically engineered tobacco and potato plants. Later, a Hepatitis B vaccine was grown in bananas. Using higher plants as production hosts offers various advantages over prokaryotic cells such as those of *E. Coli* or yeast, including lower reproduction costs, higher speed and the greater ease with which eukaryotic DNA can be manufactured.[22]

Technically, vaccines count as biologics as well, although their use predates biopharmaceuticals by centuries, with novel technologies improving their production process. Grown in chicken eggs since the 1930s, vaccine production through tissue cultures and bioreactors is more expensive yet faster and less error-prone. By infecting monkey or dog cells with a virus, for instance, large amounts of vaccine can be produced both reliably and within a short timeframe. For vaccines against measles, polio, rubella, mumps or rabies, this method is widely used.[23] For pandemic influenza, the topic of Chapter 7, it is more expensive and has not been widely adopted.

Marine organisms, which I address in Chapter 8, are increasingly of interest for the pharmaceutical sector. This includes, for instance, terpenoids synthesized via the enzymes of coral reefs. A substance with a wide range of applications, including the treatment of malaria and cancer, their biosynthesis allows for cheaper and faster production than via chemical synthesis. Chitin, extracted from crabs and lobsters, is used for its antibacterial, anti-viral and anti-fungal properties. Drugs against leukemia and viral infections have been developed from isolated compounds of the *Tethya crypta* sponge. Analgesics and anti-tumor drugs, based on isolates of cone snails, were commercialized in 2004. Presently, hundreds of marine compounds are used in the development of treatments for tumors, bacterial and fungal infections and inflammation for cardiovascular, endocrine and neurological diseases.[24]

Finally, "cosmeceuticals" is another area in which genetic resources of plant, animal and microbial origin find application. Those products are allegedly superior to ordinary cosmetics, although the absence of mandatory validation schemes makes this claim hard to substantiate. Skin care products incorporating caffeine or curcumin, a ginger molecule, are said to possess anti-inflammatory properties. Resveratrol, occurring in red grapes, is used as an antioxidant in the treatment of scarring and hyperpigmentation. *Rhodiola rosea*, from the mountainous regions of South Asia, supposedly protects against biological, environmental and psychological stressors.[25] The advantages of such products, in comparison to regular cosmetics, may well lie in the area of marketing rather than improved health benefits, though.

White biotechnology

White, or industrial, biotechnology is the most-recent commercial application. It revolves around the use of micro-organisms for fermentation processes and biocatalysis through their enzymes.[26] This allows the production of, for instance, antibiotics, pesticides or food additives, wastewater treatment or oil recovery.[27] Bio-based chemicals, excluding pharmaceutical products, had a global sales volume of EUR 48 billion in 2007. For European countries, white biotechnology is a response to its declining competitiveness in global chemicals markets relative to the surging Asian competitors.[28] The largest impacts of industrial biotechnology are expected in bioenergy. Technologies garnering significant interest in recent years are second generation cellulosic ethanol as well as fuels derived from marine micro-organisms, such as microalgae.[29] Bioenergy is likely to play a major role in mitigating climate change. The combination of bioenergy with Carbon Capture and Storage, in particular, would allow for negative emissions, as CO_2 is first sequestered in biomass and, after being used for energy generation, stored in the oceans or geologic formations in perpetuity.[30]

For the purposes of this book, industrial biotechnology in particular, and micro-organisms in general, matter differently than other, specifically plant-genetic resources. A major reason is the global distribution of micro-organisms. While the deep sea is likely to hold large reservoirs of interesting materials, microbes are often said to be "cosmopolitan," being evenly spread around the world.[31] This ameliorates the conflict between users and providers of "rare" genetic resources, which might not be found elsewhere.

Synthetic biology

Recent developments might transform the commercial application of biotechnology in fundamental ways that could, in the long run, have important implications for international ABS governance. Synthetic biology, an umbrella term often understood as the targeted, computer-assisted design of biological systems, enables the synthesis of living cells from abiotic components; genome synthesis from standardized building blocks ("biobricks"); the creation of stripped-down chassis organisms containing a minimal amount of genes, which could serve as platforms for attaching genetic "apps" with specialized functions; the synthesis of nucleic acids that do not appear in nature ("xeno nucleic acids"); the design of metabolic pathways or the directed rapid evolution of macromolecules into a desired direction.[32] Those developments overlap with the larger turn away from single genes and towards "omics," where entire genomes, proteins or metabolisms have increasingly become an object of study and commercial interest.

Those developments raise difficult questions for global governance in general, and for ABS in particular.[33] Important provisions of the Pandemic Influenza Preparedness Framework (Chapter 7) can be circumvented if vaccine viruses are synthesized from genetic sequence data, rather than derived from physical specimens.[34] Genome synthesis also complicates the task of monitoring the transboundary utilization of genetic resources as digitalized sequence data replaces the shipping of physical materials.

The relevance of genetic resources for biotechnology

For the seed industry, plant genetic resources form the raw materials for creating any new types of plant varieties, whether transgenic or not. Micro-organisms such as the *Bacillus thuringiensis* possess useful genes for creating plant varieties with properties such as insect resistance or herbicide tolerance, and *agrobacterium* is widely used for transferring foreign DNA into host plants. Microbial DNA has also been employed for imbuing plants with drought tolerance.[35] In pharmaceuticals, large amounts of drugs have traditionally been developed on the basis of plant genetic resources by directly incorporating plant DNA, harvesting proteins or synthesizing compounds that are modeled on plants. Such drugs include Aspirin®, Artemisin (a group of anti-malaria drugs), various anti-cancer drugs, Resperpine for the treatment of high blood pressure and psychosis or various drugs developed from marine genetic resources (see Chapter 8).[36] Yet since the 1990s, most of the large pharmaceutical manufacturers have slashed their natural products divisions, decreasing their reliance on plant genetic resources. Instead, they switched to the high-throughput screening of large chemical libraries for discovering useful new substances.[37] While plant and microbial biodiversity is immense and could potentially provide important leads for drug development, technological barriers to their high-throughput screening are still limiting their usefulness in contemporary pharmaceutical R&D. The present emergence of new technologies for the efficient screening of large amounts of biological materials could change this, though.[38] Things look different for one branch of pharmaceuticals, which is the focus of Chapter 7: vaccines. As I will discuss in more detail, vaccine production requires access to viral genetic resources for developing a candidate vaccine virus. Without access, vaccines cannot be produced. Particularly where highly contagious pathogens are concerned, such as pandemic influenza viruses, obtaining the relevant materials in a timely manner is of utmost importance.

For industrial biotechnology, both plant and microbial genetic resources matter. Plant genetic resources play an important role, for instance, in the production of bioenergy from corn, sugar cane, bagasse or, potentially, algae. Micro-organisms are used mainly for fermentation and biocatalysis. Yet both types of genetic resources are only of peripheral importance for questions of ABS: plant genetic resources for industrial biotechnology are rarely utilized in a transboundary context, and the "cosmopolitan" geographical distribution of micro-organisms implies that users face little problems in obtaining access. However, micro-organisms are potentially very important for the emerging ABS regime on marine genetic resources, which I discuss in Chapter 8.

Intellectual property rights

The history of modern biotechnology is inextricably linked to the history of intellectual property law. Intellectual property rights encompass a social contract. Innovation enhances public welfare, yet private actors lack the economic incentives to engage in innovation if they are unable to obtain commercial benefits therefrom. Intellectual property rights thus grant an innovator a limited period of time in which he or she may exclusively utilize an invention for commercial ends, including licensing it to third parties; simultaneously, the invention itself must be disclosed so that it may enter the public sphere after the term of protection lapses.

Two types of intellectual property rights are relevant for this book: patents and plant variety rights. Patents are the strongest form of protection an innovator can receive. While patent laws vary among countries, sometimes significantly so, the consent of the right holder is generally required for the production, use, sale or import of patented products and processes within the jurisdiction where protection has been granted. As are all other forms of intellectual property rights, patents are territorial in nature. While national patent laws are increasingly being harmonized through international treaties, and while mechanisms exist for obtaining a bundle of patents in multiple jurisdictions through a single application, the protection conferred within one jurisdiction does not automatically extend to other jurisdictions.

Plant variety rights have emerged as an alternative regime towards the middle of the 20th century. They are geared towards the specific nature of agricultural breeding. For producing new varieties of plants, breeders typically require dozens, or even hundreds, of existing varieties, which undergo several cycles of cross-breeding until a variety with certain desirable properties emerges. Protecting plant varieties with patents hampers this process by requiring the breeder to obtain the consent of each and every right holder for each and every variety used in the breeding process. While plant variety rights also differ among countries, they allow for certain unauthorized uses by third parties, for instance as input materials for developing new varieties.

Both patents and plant variety rights have undergone significant changes since the latter half of the 20th century, particularly since the 1980s. Those changes have played out both on the national and the international level. In industrialized countries, which have a significantly higher rate of innovation than others, evolving case law has expanded the scope of patentable subject matter for genetically modified organisms as well as the processes used for their production. Internationally, the 1995 TRIPS agreement proscribes minimum standards for WTO members' domestic patent laws. Virtually all Free Trade Agreements contain provisions on intellectual property rights, which often go far beyond what is required under TRIPS. Likewise, UPOV provides model legislation for plant variety rights contracting parties can transpose into their respective legal systems; yet protection standards under UPOV have been ratcheted up through repeated revisions of the treaty text. The international diffusion of UPOV standards, including to developing countries, similarly takes place through trade agreements.

I first discuss changes in US and European law, which have expanded the scope of protectable subject matter and strengthened private ownership over plant varieties and inventions incorporating genetic resources. I then turn to international treaties setting minimum standards for intellectual property law and the ways in which large parts of the world have been made to accede to those treaties.

The US

The foundational event in the emergence of commercial biotechnology was the US Supreme Court's 1980 landmark decision in *Diamond v. Chakrabarty,* revolving around a patent claim on an oil-eating bacterium engineered to clean up oil spills. Until then, living things were not patentable subject matter under US law. In *Diamond*, the Supreme Court held that the bacterium constituted "a nonnaturally occurring manufacture or composition of matter," the question of whether or not it is a living organism being irrelevant to the question of patentability.[39] Subsequent case law expanded the scope of patentable subject matter to transgenic plants. Weaker protection than that granted through regular utility patents had already existed for plants under the 1930 Plant Patent Act (for asexually reproducing varieties) and the 1970 Plant Variety Protection Act (for sexually reproducing ones). In 1985, the US Patent and Trademark Office affirmed the application of the *Diamond* decision to transgenic plants, seeds and tissue cultures in *ex parte Hibberd*. Regular utility patents became available for sexually reproducing plants, including non-transgenic ones, through the Supreme Court's 2001 decision in *J. E. M. Ag Supply v. Pioneer Hi-Bred International.*[40] As the court argued, the existence of the 1970 Plant Variety Protection Act was not intended as an exclusive protection regime. Breeders could thus seek out other forms of protection, including utility patents, providing the applicable legal criteria were met. The Court expanded the scope of patent protection once more in 2013 when it held, in *Bowman v. Monsanto*, that patent protection on seeds extends to its progeny. Replanting saved seeds from patented varieties thus required the consent of the right holder.

For animals, the story is similar. In the wake of *Diamond*, the US Patent and Trademark Office affirmed that animals may be patented as well. In 1987, the office changed its policy to henceforth consider "non-naturally occurring non-human multicellular living organisms, including animals, to be patentable subject matter."[41] In 1988, the first such patent was granted for the "Oncomouse" developed by Harvard University, a genetically modified mouse highly susceptible to cancer, thus having useful applications for drug development. Thousands of animal patents have been filed since and, despite the significant social and political controversies, none of the various legislative proposals to ban such patent claims successfully passed the US Congress.[42]

What about humans? As of today, humans as such are not patentable, neither in the US nor elsewhere. But things are different for isolated human genes. In the 1990s, the Human Genome Project set off a bonanza of such patent claims. Patents were even granted in instances where the only inventive activity included the isolation of the gene and the demonstration of the usefulness of its particular function. Under this permissive interpretation of patentability criteria, the US Patent and Trademark Office granted patents on the *BRCA* 1 and 2 genes in the mid-1990s. If those genes are damaged, the risk of ovarian- and breast cancer increases significantly. Patent protection allowed the right holder, *Myriad Genetics*, to charge monopoly prices for a cancer screening test that, on the free market, would have cost a fraction. In a rare instance of turning back the wheel, the US Supreme Court ruled in 2013 (*Molecular Pathology v. Myriad Genetics*) that the mere isolation of a gene is insufficient for obtaining patent protection, unlike in the case of *Diamond* in which the patent claim applied to biological materials that had undergone actual biotechnological innovation.[43]

Europe

Developments in Europe were similar.[44] The main legislative instruments addressing biological patents are the 1963 Strasbourg Convention on the Unification of Certain Points of Substantive Law on Patents for Invention; the European Patent Convention (EPC) of 1973; and the EU's 1998 Biotechnology Directive. The EPC covers all of the member states of the EU as well as several non-EU states such as Norway, Switzerland and Turkey. The EPC requires its contracting parties to grant patents on inventions that are new, involve an inventive step and are capable of industrial application, yet does *not* allow patents on, *inter alia*, "plant or animal varieties or essentially biological processes for the production of plants or animals"; this exemption does not cover "microbiological processes or the products thereof."[45] The same provisions are found in the EU's 1998 Biotechnology Directive, which was to bring EU law in the area of biopatents in line with the TRIPS agreement (see below). The EU also possesses a system for the protection plant varieties under the 1994 Regulation on Community Plant Variety Rights. Unlike for patents, those rights cover the entire Union instead of individual member states.

The expansion of the scope of patentable subject matter took place via the successive narrowing of the EPC's Article 53(b) exemptions from patentability for plant variety rights and essentially biological processes for the production of plants. European law understands "plants" as a broad category, which encompasses "plant *varieties*" in a non-exhaustive manner. Under the 1994 Community Plant Variety Rights regulation, a "plant variety" is "a plant grouping within a single botanical taxon of the lowest known rank," which can be defined via the expressions resulting from its genotype or genotypes, distinguishes itself from other plant groupings and is stable in its reproduction.[46] This echoes the provisions of the UPOV Convention, which requires plant varieties to be distinct, uniform and stable. Conversely, the European Patent Office (EPO) understands "plants" to be "an abstract and open definition embracing an indefinite number of individual entities defined by a part of its genome or by a property bestowed on it by that part."[47]

While plant varieties are thus not patentable pursuant to article 53(b) EPC, plants are. Since the EPO's 1988 ruling in *Lubrizol*, EPC members must grant patents on plants that are not plant varieties, that is, are not stable, distinct and/ or uniform. This allowed patents on hybrid seeds that do not transmit their hereditary information in a stable manner to their progeny; not being stable, they are not considered plant varieties and thus escape the article 53(b) exemption. Plant cells, similarly, have been patentable since the EPO's 1995 decision in *Plant Genetic Systems*.

The EU's 1998 biotechnology allows indirect patent claims that pertain not to one, but to multiple plant varieties: biotechnological inventions are patentable "if the technical feasibility of the invention is not confined to a particular plant [...] variety."[48] Thus, while plant varieties may not be claimed *individually*, multiple varieties may be claimed *jointly* if the scope of the respective invention is sufficiently broad. Patent claims to the cells of a particular plant variety are equally permissible. This reasoning was affirmed by the EPO's 1999 *Novartis II* decision, arguing that a patent claim's subject matter may differ from its scope: patents are thus admissible for inventions that "embrace" plant varieties without, however, individually claiming them.[49]

Similar to that of plant varieties, the scope of the exemption from patentability of "essentially biological processes" has been whittled down. This exemption was intended to prevent traditional breeding methods from coming under exclusionary rights. Yet, what exactly makes a process *essentially* biological? The EPO originally held that the quality of human intervention "had to be decisive in determining whether a process was biological in its essence or not"[50] and thus whether or not it would constitute patentable subject matter. In *Lubrizol*, the EPO argued that this exemption had to be "narrowly construed"; whether or not a process would fall within the scope of article 53(b) had to be "judged on the basis of the essence of the invention taking into account the totality of human intervention and its impact on the result achieved."[51] Even where a process consists of steps that are essentially biological individually, their interaction and specific sequence can lead to the process ceasing to be "essentially" biological. In *Plant Genetic Systems* (1995), the EPO ruled that the existence of "at least one essential technical step, which cannot be carried out without human intervention and which has a decisive impact on the final result" is sufficient for a process to become non-essentially biological, thus escaping the process exclusion.[52] Under the Biotechnology Directive, a process is understood to be essentially biological "if it consists *entirely* of natural phenomena such as crossing or selection."[53]

The EPO decided, in *Tomato 1* and *Broccoli 1*, that a process is patentable if it comprises a technical step that "introduces a trait into the genome or modifies a trait in the genome of the plant produced, so that the introduction or modification of that trait is not the result of the mixing of the genes of the plants chosen for sexual crossing,"[54] with processes for the production of transgenic plants accordingly constituting patentable subject matter. In *Tomato II* and *Broccoli II* of 2015, cases that dealt with patent claims to the *processes* for producing a tomato with reduced water content and a broccoli with increased glucosinolates as well as the *products* themselves, the EPO ruled that plants (but not plant varieties!) are patentable even if the processes for their production, being essentially biological, are not. In other words: the process exclusion was understood to not extend to the products resulting from such processes, yet even essentially biological processes may indirectly be patented through product-by-process claims.

International minimum standards

In Europe and the US, the two regions that account for the lion's share of global biotechnological innovation, change was endogenous, with inventors trying to obtain favorable judicial interpretations by pushing the envelope of patent law. In other parts of the world, this is different: in order to protect their intellectual property abroad, industrialized countries have been pushing for strict international minimum standards since the 1980s. Over time, this led to a diverse landscape of international treaties dealing with patents and plant variety rights.[55]

To start with the latter, the 1961 International Convention for the Protection of New Varieties of Plants (UPOV in the French acronym) provides a model for national legislation to protect the intellectual property of plant breeders. With numerous plant varieties serving as input materials for the breeding of new plant varieties, patent protection may not be feasible. UPOV attempted to create a regime that balances the commercial interest of breeders without stifling third-party access to input materials. Plant variety rights are granted for varieties that are distinct, stable and uniform. Unlike patents, they allow for several types of unauthorized use, the Breeder's Exemption. Protected varieties can thus be used for all purposes within the scope of this exemption, without the authorization of the right holder.

Analogous to the legal changes discussed above, UPOV members have reduced the scope of this exemption over time. In the 1961 version of UPOV, the right holder must authorize the production, commercial marketing or offering for sale of protected varieties.[56] In UPOV 1991, authorization is required for the production, conditioning, offering for sale, selling or marketing, exporting or importing, and stocking for any of those purposes.[57] Between the 1961 and the 1991 version, terms of protection were extended from a minimum of 15 to a minimum of 20 years. UPOV 1991 also extends to materials harvested from a protected variety. The right to reuse farm-saved seeds is known as the "farmers' privilege." Finally, the 1991 version also restricts the "farmers' privilege" of re-planting seeds harvested from protected plant varieties. UPOV members may allow farmers to reuse such seeds "on their own holdings" yet only "within reasonable limits" and "subject to the safeguarding of the legitimate interests of the breeder,"[58] which may encompass monetary payments.[59]

The changes in UPOV protection standards "represent a huge step in the direction of the patent system."[60] Yet patent standards have, themselves, been pushed in an upwards direction. The 1995 Agreement on Trade-Related Aspects of Intellectual Property Rights (TRIPS) is compulsory for WTO members, requiring them to grant patents on all inventions, including biotechnological ones, given that they are new, involve an inventive step and are capable of industrial application.[61] Members may exclude from patentability inventions that are contrary to morality or *ordre public,* "diagnostic, therapeutic and surgical methods for the treatment of humans or animals" and "plants and animals other than micro-organisms" as well as "essentially biological processes for the production of plants or animals other than non-biological and microbiological processes."[62] Most WTO members have used the *ordre public* exemption to ban patents on cloned humans or their genetic materials. Unconditionally exempting plant (and animals) from patentability under *ordre public* is generally not considered permissible; where TRIPS members exempt plants, they must provide for an "effective *sui generis* system" for the protection of plant varieties.[63] For many actors, including the secretariats of WTO and UPOV, the international plant breeders' association ASSINSEL and various industrialized countries, such a system is provided by UPOV,[64] yet the extent to which TRIPS members enjoy flexibility in drafting their own *sui generis* regimes is a matter of dispute. Finally, the possible exemption of "diagnostic, therapeutic and surgical methods for the treatment of humans or animals" is relevant for pharmaceutical biotechnology yet not agricultural one.

UPOV was originally an exclusive club of (mainly) industrialized countries. As developing countries came under pressure to fulfill their article 27.3(b) obligations under TRIPS in the 1990s, membership spiked, from 20 in 1992 to 72 in 2016.[65] Accession to UPOV is also mandatory under various Free Trade Agreements the EU, Japan and the US have concluded with third countries. The US-Chile agreement, for instance, obliges its parties (that is, Chile) to adopt the 1991 version of UPOV as well as "undertake reasonable efforts" to enable patent protection for plants. Agreements with Colombia, Costa Rica, El Salvador, Guatemala, Honduras, Nicaragua, Peru, Panama and the Dominican Republic contain similar language. The EU has concluded such treaties with countries such as Algeria, Colombia, Jordan, Lebanon, Morocco, Peru and Syria, making the adoption of UPOV 1991 mandatory and, in some cases, requiring patents to be granted on transgenic crops. Japan has such treaties in place with Brunei, Chile, Indonesia, Malaysia, Thailand and Vietnam, among others.

Folding those requirements into larger package deals is necessary to overcome the skepticism of developing countries, as net importers of intellectual property, towards implementing legal changes that might hamper domestic innovation and merely protect the commercial interests of foreign investors.[66] Even authors who are cautiously optimistic about the benefits of plant variety rights in developing countries warn about the risks of excessively stringent, one-size-fits-all protection regimes.[67] And where transgenic crops are concerned, "the instruments that are being used for its protection have become highly exclusionist in their approach" and "seem to foreclose the entry of latecomers into the technology race."[68] Ownership of relevant technologies remains highly concentrated, indeed. This is where I turn next.

Industry structure and evolution

Industry has quickly embraced the biotechnological revolution, yet different sectors have done so in different ways. Most of today's global players in agricultural biotechnology have emerged from chemicals companies such as Dupont, Monsanto and Dow in the US; Hoechst, BASF and Bayer in Germany or Ciba and Geigy in Switzerland. Having built their industrial bases in agrochemicals and industrial chemicals, they rapidly began adopting biotechnology in the 1980s.[69] In Europe, large chemicals companies began branching out into pharmaceuticals at about the same time, with Bayer, Hoechst or Ciba-Geigy (merged in 1971) and F. Hoffmann-La Roche moving into molecular biology and biotechnology in the 1970s and 1980s. In the US, agriculture and pharmaceuticals largely went separate ways, with companies such as Merck, Abbott, Schering-Plough and Eli Lilly branching out into areas such as recombinant vaccines, monoclonal antibodies and interferon since the 1970s.

Those industrial structures changed profoundly in the 1990s with the rise (and later fall) of large life sciences conglomerates. At first, mergers and acquisitions on an unprecedented scale were driven by the attempt "to leverage new biotechnology techniques across numerous industry sectors."[70] Sandoz and Ciba-Geigy merged their agricultural and pharmaceuticals divisions to Novartis in 1996; in 1999, Monsanto merged with Pharmacia & Upjohn, Hoechst with Rhône-Poulenc (forming Aventis), and Astra with Zeneca (to AstraZeneca). Yet the gold rush did not last. Due to the public and regulatory backlash against agricultural biotechnology and genetically modified food in the late 1990s, "agricultural biotech firms were keen to separate out the different social, political and economic risks involved in biotechnology and to shield medical applications from the negative publicity that GM food attracted."[71] This risk-separation strategy led to the demerging of pharmaceutical and agricultural companies:[72] in 2000, AstraZeneca and Novartis spun off their agriculture operations and formed Syngenta. Monsanto and Pharmacia went their separate ways in 2000. Bayer acquired Aventis' agricultural division in 2002 and formed Bayer CropScience. Yet within the agricultural sector, mergers and acquisitions continue to the present day. The US $130 billion merger between Dow Chemical and DuPont, announced in December 2015, is the largest merger in the industry yet. At the time of writing, the state-owned Chinese company ChemChina has bid US $43 billion for a takeover of Syngenta, and Bayer AG is considering a bid for Monsanto.

Since the 1990s, the agricultural and pharmaceuticals sectors have evolved along different paths. In the former, consolidation led to highly integrated global players active in all areas of agricultural inputs. One study estimates that three companies control over 50% of the global seeds market (Monsanto, Dupont Pioneer and Syngenta) and over 50% of the global pesticides market (Syngenta, Bayer CropScience and BASF).[73] Those dominant positions result from not only mega-mergers, but also the acquisition of medium-sized, specialized companies, including from the Global South: since the mid-1990s, Monsanto bought the biotech firms Agracetus and Calgenet the seed producers Holdens Foundation Seeds, DeKalb, Delta and Pineland and parts of Cargill and Asgrow.[74] DuPont acquired Pioneer Hi-Bred in 1998. More recently, Syngenta bought up the Italian seed company Società Produttori Sementi, and the French seed giant Vilmorin acquired parts of Campbell Soup, of Andy Warhol fame. In Africa, Monsanto, Syngenta, Dupont Pioneer and the French Groupe Limagrain bought up parts of the continent's largest seed companies, SeedCo, MRI Seed and the South African companies Carnia, Sensako and Pannar Seed.

Corporate concentration in agriculture is partially a result of genetic transformation technology, plant varieties and genes being complementary assets, creating incentives for companies to gain control of all three.[75] The entry of large agrochemicals companies into biotechnology also resulted from expiring patent terms on pesticides in the late 1990s: where transgenic crops are protected with patents, licensing agreements can prescribe the pesticides buyers must use, allowing the marketing of profitable "crop-plus-pesticide" packages. Somewhat ironically, concentration was also driven by the strict regulations several countries (and the EU) put in place during the 1990s, with the costs of regulatory compliance creating market entry barriers for smaller newcomers

At the same time the private sector surged, the public one declined due to a stagnation of funding and a wave of privatizations.[76] As one study puts it, "[p]ublic sector capacity in the agricultural sciences in many countries has been run down over decades, infrastructure has been depreciated and the majority of the scientists in many countries are close to retirement age."[77] Notable exceptions are China, India and Brazil. Here, total public sector spending on agricultural R&D is comparable to that in the top-3 high-income countries (US, Japan and Germany).[78] The former are also heavily investing in the development of transgenic crops and hold significant *ex situ* collections of plant germplasm in public collections.[79] Other exceptions are the programs of international agricultural research centers, such as the Mexican Centro Internacional de Mejoramiento de Maíz y Trigo (CIMMYT) and the International Rice Research Center in the Philippines, which were at the forefront of crop improvement and significantly contributed to the Green Revolution towards the middle of the 20th century.[80] In 1971, they were among the founding members of the Consultative Group for International Agricultural Research (CGIAR), a consortium of presently 15 research centers that has had a profound impact on crop improvement.[81] The CGIAR is, today, the largest holder of *ex situ* samples of Plant Genetic Resources for Food and Agriculture and thus figured prominently in the negotiations on the FAO Seed Treaty (see Chapter 5).

In pharmaceuticals, the story is slightly different. Several high-volume mergers and acquisitions have characterized the industry since the late 1980s, with mergers between Bristol Myers and Squibb (1989); SmithKline Beckman and Beecham (1990); Glaxo and SmithKline Beecham (1995); and acquisitions of Marion Merrel Dow, a Dow subsidiary, by Hoechst (1995), of Genentech by F. Hoffmann-La Roche (1990), Wellcome by Glaxo (1995) or Syntex by Roche (1994 to 1996). At the same time, pharmaceutical companies have entered into a wide range of strategic alliances, usually between 30 and 40 each, with the numerous small and medium-sized biotechnology companies entering the market since the 1980s.[82] This led to vertical disintegration. The small and medium-sized players possessing highly specialized knowledge while lacking the capacities for long-term product development, including clinical trials and obtaining regulatory approval. From the mid-1990s they began acting as a supporting nexus,[83] selling potential drug compounds and services to those large pharmaceutical companies downstream that are "prepared to spend a quarter of their R&D budgets externally to ensure a smooth supply of new products."[84] Universities have also begun playing a crucial role within the larger biotech-pharma network, with a new generation of "gene jockeys" straddling the border of non-profit, basic research and applied research for commercial ends.[85]

The vaccine sector, of crucial importance for the case study in Chapter 7, only accounts for a minor fraction of the global pharmaceutical market: 3% as of 2012. With large numbers of manufacturers exiting the market since the 1970s, 6 companies (Merck, GlaxoSmithKline, Sanofi Pasteur, Sanofi Pasteur MSD, Pfizer, Novartis) presently account for 80% of global production. Those are chiefly located in Europe.[86] Compared to small-molecule drugs, vaccine production "is less standardized and less predictable," involving "the complex transformation of live biologic organisms into pure, active, safe, and stable immunization components."[87] Several large pharmaceuticals manufacturers have recently divested from vaccines due to a lack of profitability, frequently withdrawing vaccines already approved for marketing.[88]

The broader picture

The changes sketched out above have led to biotechnology's becoming an increasingly central feature of the world economy. How large the economic relevance of (different types of) genetic resources is across different industrial sectors cannot be said with certainty and arguably undergoes historical fluctuations. When the discussion on how to regulate the commercial utilization of genetic resources commenced in the 1980s, the expectations were that incredibly valuable materials were being used, often illegitimately, by a handful of companies in the Global North while the rest of the world was excluded from the genetic bonanza. From today's perspective, those expectations appear exaggerated yet are in line with the often rather naive optimism regarding the economic potential of biotechnology more broadly.

At the same time, genetic resources and ABS are linked to broader issues beyond the narrow question of who gets what. The transformation of the global seed industry has raised various concerns regarding the increased usage of genetically modified seeds and the social and environmental risks this entails.[89] The international diffusion of regimes for plant variety protection has profound effects on seed systems in developing countries, potentially altering the ways in which plant breeding has been carried out for millennia.[90] The recognition of the potential economic value of genetic resources takes place parallel to a loss of biological diversity, which in its scale and speed has probably occurred a handful of times in the entire history of the planet and has led to attempts to precisely measure and quantify the economic benefits of biological diversity more broadly.[91] For tackling climate change, the production of bioenergy on a massive scale, possibly in conjunction with technologies for Carbon Capture and Storage, appears more and more unavoidable,[92] a development with probably profound impacts on both the commercial value of biomass and attempts to conserve biological diversity.[93] Gene drives offer a novel option for the biological control of pests and invasive alien species yet might pose significant hazards for biodiversity.[94] Things look similar for the oceans: at the same time that stakeholders are beginning to realize the value of genetic resources in the deep sea and international waters,[95] marine ecosystems are under immense pressure from overfishing, pollution, waste, mining and acidification. Thus, genetic raw materials of potentially immense value for pharmaceutical product development might be lost forever. For public health more broadly, the shrinking growth potential of chemical engineering implies that combating new and emerging health threats might, in the long run, require a much stronger emphasis on biotechnology, including product development based on microbes or plant materials.

Genetic resources are intrinsically connected to sustainable development. It is thus an unfortunate but probably unavoidable fact of life that, in their governance, national egoisms usually prevail over a vision of the common good: instead of making plant genetic resources for food and agriculture available free of charge in order to improve global plant breeding, important provider countries keep tabs on their most-valuable materials in the hopes of extracting profits for themselves.[96] Major industrialized countries have stalled attempts to equitably share pandemic influenza vaccines with developing countries despite the latter's providing large shares of the viral specimens on the basis of which those vaccines are developed.[97] Some of the (very few) countries capable of extracting microbes and other marine organisms from the oceans and subsequently using them in commercial applications consider it perfectly legitimate to realize gains for themselves while shutting everyone else out.[98] Other countries make a habit of styling themselves as the champions of the Global South in ABS governance yet abandon ship when it is expedient to do so.[99]

In the chapters that follow, I do not address the question of how far fairness, equity, common concerns and responsibilities and other concepts that form the normative basis of ABS governance are genuine motivations for governments' bargaining behavior and how far they are simply rhetorical devices for enhancing individual gains from international cooperation.[100] This especially applies when we look at the shifting normative underpinnings of ABS: the question of who has which rights of ownership over genetic resources. This is where I turn next.

Notes

1 Pujar et al. (2015), 3.
2 Convention on Biological Diversity, Article 2.
3 OECD (2005), 9.
4 Convention on Biological Diversity, Article 2.
5 While falling under the CBD's definition in Article 2, human genetic resources have subsequently been excluded due to their ethically controversial nature. See CBD COP 2, Decision II/11.
6 Tvedt and Schei (2014), 20–21.
7 Tvedt and Schei (2014).
8 Broggiato et al. (2014), 177.
9 Murphy (2007b).
10 NAS (2016b), 44–45.
11 NAS (2016b), 45.
12 Kinchy (2012).
13 NAS (2016b), 45.
14 Qaim and Zilberman (2003).
15 Parisi et al. (2016), 32.
16 Qaim and Zilberman (2003); Brookes and Barfoot (2008); Gusta et al. (2011); Barrows et al. (2014).
17 Mann (2010), 149.
18 Gilbert (2013).
19 Smith Hughes (2012).
20 Rasmussen (2014), 68 and 102.
21 Knäblein (2013), 20–21.
22 Warzecha (2012).
23 Pujar et al. (2015), 5.
24 OECD (2013), 31–32.
25 For an overview, see Sivamani et al. (2016) and Gurib-Fakim (2014).
26 Soetaert and Vandamme (2005); molecules derived from plants or animals play a lesser role here than microbial ones.
27 Harzevili (2015); for a broader overview, see also Pandey et al. (2015).
28 OECD (2011), 51; critically Richardson (2012).
29 Carvalho and Pereira (2015).
30 Azar et al. (2010); critically Fuss et al. (2014).
31 Green and Bohannan (2006).
32 Presidential Commission (2010); CBD (2015), 17–21; Hutchison III et al. (2016); ETC Group (2016).
33 CBD (2015).
34 Gostin et al. (2014).
35 NAS (2016b), 50.
36 Pearce and Puroshothaman (1995); Sampath (2005), 13.
37 Sampath (2005).
38 Harvey et al. (2015).
39 Palombi (2009).
40 Rimmer (2008), 59–65.
41 USPTO, Animals - Patentability, 1077 O.G. 24, April 21, 1987.

42 Koepsell (2015), 74.
43 Koepsell (2015), 89–100.
44 For a comprehensive overview of EU intellectual property law, see Kur and Dreier (2013).
45 EPC 1973, Article 53(b).
46 Council Regulation (EC) No 2100/94, Article 5.2.
47 G 0001/98, 17.
48 Biotech Directive Article 4.2; see also Leskien (1998).
49 EPO Decision of 20 December 1999. Case number G 0001/98.
50 EPO, Decision of the Technical Board of Appeal 3.3.2. of 10 November 1988, 1.
51 EPO, Decision of the Technical Board of Appeal 3.3.2. of 10 November 1988, 8–9.
52 EPO, Boards of Appeal, Decision of February 1995, case number T 0356/93 – 3.3.4., 36.
53 Directive 98/44/EC of the European Parliament and of the Council of 6 July 1998 on the legal protection of biotechnological inventions. Article 2.2; my italics.
54 G 0002/07, 69.
55 Helfer (2009); Rabitz (2014).
56 UPOV 1961, article 5(1).
57 UPOV 1991, article 14(1).
58 UPOV 1991, article 15(2).
59 Dutfield (2011), 9; see also Srinisavan (2010).
60 Andersen (2008), 158.
61 TRIPS, Article 27.1.
62 TRIPS, Article 27.3.
63 TRIPS, Article 27.3(b).
64 Dhar (2002).
65 This excludes regional organizations.
66 Helfer (2009); Muzaka (2011).
67 Tripp et al. (2007).
68 Chaturvedi (2009), 369.
69 Chandler (2005).
70 King et al. (2002), 14.
71 Falkner (2009), 144; see also Joly and Lemarié (1998).
72 Fulton and Giannakas (2001).
73 ETC Group (2013); see also Howard (2009).
74 Brennan et al. (2000), 154.
75 Graff et al. (2001).
76 Eicher and Rukuni (2003); Heisey et al. (2001).
77 Pardey et al. (2013), 111.
78 Pardey et al. (2013), 110.
79 Parisi et al. (2016), 33–34.
80 Murphy (2007b), 90–94.
81 Renkow (2010), 391–402.
82 de Rond (2003), 39.
83 Chandler (2005); see also Cockburn (2004); Grabowski and Vernon (1994); Schweizer (2005).
84 Nightingale and Mahdi (2006), 76; see also Dosi and Mazzucato (2006), 4–5; Cefis et al. (2006).
85 Rasmussen (2014).
86 PhRMA (2013), 45; EVM (2004).
87 NAS (2003), 109.
88 NAS (2003), 125–126.
89 Baram and Bourrier (2011).

90 Kloppenburg (2004); Murphy (2007a); Tripp et al. (2007).
91 Atkinson et al. (2014); Helm and Hepburn (2014).
92 Azar et al. (2010).
93 ETC Group (2010).
94 NAS (2016a).
95 OECD (2013).
96 Moore and Tymowski (2005).
97 Kamradt-Scott and Lee (2011).
98 Arnaud-Haond et al. (2011).
99 Nijar 2011.
100 On the link to justice, see Coolsaet and Pitseys (2015).

4   Property regimes

Underlying the global governance of genetic resources are normative conflicts regarding ownership and control: should their private ownership be permissible, and which obligations should owners have? Are genetic resources the common heritage of mankind, under the control of sovereign nation states or the common concern of all countries? What about the rights of farming communities or local and indigenous peoples who have cultivated and conserved genetic resources for significant periods of time? And to what extent should different types of property regimes apply to different types of genetic resources? Should resources used for plant breeding fall under the same (or a similar) regime as resources used for the development of pharmaceutical products? And what about genetic resources in areas beyond national jurisdiction?

While this book takes a straightforward institutionalist approach to the global governance of ABS, those normative questions are at the core of international debates on the design of appropriate rule systems. The inherent difficulty of finding common ground on moral issues once relational goods are involved is further complicated by a lack of objective knowledge regarding key factual aspects relevant to ABS, such as: what is the precise economic relevance of genetic resources for biotechnology? Are patents and plant variety rights the best way to stimulate their socially beneficial utilization? To what extent are genetic resources being illegally appropriated and used? And what are the resulting opportunity costs to providers? What is the most effective way of preventing misappropriation and misuse?[1] Finally, there is the question of whether the motives of (some) developing countries in creating strong international ABS regimes are sincere: is fair and equitable benefit-sharing and/or the prevention of "biopiracy" really the major motivating factor, or is the issue being used instrumentally as part of a hidden agenda to soften up the international patent regime, as some industrialized countries allege?[2]

The topic of this chapter is how the former set of questions has guided institutional and legal changes and spurred political conflicts among different groups of actors. I focus on the common heritage approach to genetic resources, private property rights and the state sovereignty paradigm. The ABS regimes to be discussed in subsequent chapters usually mix several of those paradigms in ways that have sometimes led to rule inconsistencies. Yet the discussion of the three paradigmatic approaches also prepares the ground for explaining negotiation behavior. Each of the three has specific implications for the distribution of gains from international ABS regimes, and each is being deployed as part of negotiators' rhetorical action for pushing the international debate in specific directions.

Intellectual property rights and the public interest

While private ownership of biological materials is controversial, it is important to recall its underlying reasons. Plant breeding and pharmaceutical innovations require large investments over longtime horizons with uncertain outcomes. Bringing a new drug from the laboratory to the market, including clinical trials and regulatory approval, can take upwards of 10 years and US $1 billion.[3] Both the money and time required has increased over the last decades as the "low-hanging fruit" of small-molecule drugs has been plucked and manufacturers are moving into large-scale molecules, which are more difficult to develop and involve larger uncertainties.[4] Things look similar for plant breeding. Breeders must generate sufficient genetic variation through crossing, induced mutagenesis or genetic engineering, select the most suitable plants and test them in the field.[5] For transgenic plants, regulatory compliance poses additional hurdles. In addition to the economic risks, small and medium-sized biotech companies working upstream also face difficulties accessing finance.

The political controversies over intellectual property rights sometimes obscure their original justification, which in one line of reasoning is a social contract aimed at balancing the public interest with private profit-seeking behavior:[6] societies collectively require innovation for enhancing their welfare yet inventors may lack the economic incentives to invest large sums of money into uncertain R&D processes if they do not expect adequate profits. Intellectual property rights attempt to balance those competing interests: a time-limited, artificial monopoly is granted on inventions (or new plant varieties) in exchange for the disclosure of the ways to produce them. Once protection lapses, the disclosed information enters the public sphere and is freely usable by everyone. In theory, both sides benefit: "the private party can recoup the investment in research and development, whilst the public acquires knowledge of a new invention and can make use of it without extra investment" once protection expires.[7]

The purpose of intellectual property rights is thus to enhance social welfare by encouraging innovation in the long term. This has to be balanced against the costs of creating artificial monopolies and barriers to inventions that could be made publicly available in the present if intellectual property protection did not exist. Those competing values are most apparent in the debate over access to medicines: artificial scarcity makes many life-saving drugs unavailable for people who have an objective medical need; yet, as the counter-argument goes, infringing on the property rights of pharmaceuticals manufacturers would jeopardize the long-term viability of creating newer and better drugs.

Intellectual property rights can also have a deeper, transformative impact on knowledge systems. The "anti-commons" effect refers to the hampering of innovation by limiting the capacity of others to draw upon protected knowledge for their own innovative ends, a hypothesis for which there is some empirical evidence.[8] The wider social impact of the biotechnological anti-commons is especially pronounced as it emerged in an area that had traditionally been governed not by market relations, but rather by the free exchange of the input materials for those innovation processes that arguably laid the foundations of human civilization and where the distribution of benefits resulting from the enclosure of the commons is highly asymmetrical.

Common heritage and the seed wars

The idea of common property reaches back over the philosophy of enlightenment, early Christian thought and Roman law up to the Socratics. Termed the "common heritage of mankind," it first became a principle of international law in the 20th century. The 1954 Hague Convention for the Protection of Cultural Property in the Event of Armed Conflict seeks to safeguard "movable or immovable property of great importance to the cultural heritage of every people." The 1967 Outer Space Treaty provides that outer space shall be used and explored "for the benefit and in the interests of all countries" and is "not subject to national appropriation by claim of sovereignty." The 1959 Antarctic Treaty invokes the principle indirectly by stating that "it is in the interest of all mankind that Antarctica shall continue forever to be used exclusively for peaceful purposes and shall not become the scene or object of international discord." And the 1982 United Nations Convention on the Law of the Sea declares the seabed, the ocean floor and its subsoil ("the Area") as the common heritage of mankind where "[n]o State shall claim or exercise sovereignty or sovereign rights over any part of the Area or its resources, nor shall any State or natural or juridical person appropriate any part thereof."[9]

The common heritage principle was at the core of the 1983 International Undertaking on Plant Genetic Resources. The Undertaking was "to ensure that plant genetic resources of economic and/or social interest, particularly for agriculture, will be explored, preserved, evaluated and made available for plant breeding and scientific purposes" based on "the universally accepted principle that plant genetic resources are a heritage of mankind and consequently should be available without restriction."[10] When the Undertaking was adopted, agricultural biotechnology was gaining traction and commercial breeders began claiming larger and larger amounts of intellectual property over seeds. Yet which genetic resources, specifically, should be available "without restriction"? The Undertaking referred both to "raw" plant genetic resources (land races, wild and weed species) and improved plant varieties, including "special genetic stocks."[11] While a more detailed discussion has to wait until the following chapter, this opened a fault line between "those in favor of intellectual property rights over improved varieties of plants, and those in favor of unrestricted access to all plant varieties."[12]

Plant genetic resources are a special type of resource. From the Neolithic revolution onward, the unhindered exchange of plants and seeds has been the basis on which human agriculture developed and thrived. By combining genetic inputs from all parts of the world, and specifically from the centers of origin of plant biodiversity, farmers were able to create new and better plants, which could subsequently be used for breeding by others.[13] For the better part of the 12,000 or so years since the dawn of agriculture and human civilization, seeds were not subject to private property rights. This only began to change in the latter half of the 20th century—when policy makers became increasingly aware that Earth's biodiversity, including the crop genetic resources cultivated over centuries and millennia, is eroding at an alarming rate. The International Undertaking thus not only aimed at the question of private ownership over seeds, but also at the conservation and sustainable use of mankind's common heritage.

The Undertaking was part of wider social and political struggles: the "seed wars" (or sometimes "gene wars") of the 1980s.[14] The seed wars saw a broad array of civil society organizations, developing countries and formidable intellectuals and activists such as Cary Fowler, Pat Mooney and Vandana Shiva, stepping up against agricultural biotechnology, corporate control over seeds and agricultural inputs and the wider social and technological changes associated with the transformation of global agricultural systems. As Shiva, an Indian activist, notes in her criticism of global agriculture's Green Revolution of the 1950s and 1960s, "[f]or 10,000 years, farmers and peasants had produced their own seeds, on their own land, selecting the best seeds, storing them, replanting them, and letting nature take its course in the renewal and enrichment of life. With the Green Revolution, peasants were no longer to be custodians of the common genetic heritage through the storage and preservation of grain. The 'miracle seeds' of the Green Revolution transformed this common genetic heritage into private property, protected by patents and intellectual property rights."[15] Despite the tendency towards romanticizing the autonomy farmers allegedly enjoyed prior to the Green Revolution and the subsequent emergence of agricultural biotechnology, Shiva correctly pointed out the monumental, historical changes that were taking place in agriculture since the 20th century. The Canadian activist Pat Mooney, commenting on the loss of agricultural biodiversity as a consequence of mono-cropping and the broader industrialization of agriculture, noted in 1983 that "[i]n the space of a few decades, the creation of high-yielding crop varieties wiped away the little genetic diversity that did exist in the North" yet "the so-called 'breadbasket' nations may be grain-rich but they are gene-poor and wholly dependent upon the Third World for the long-term survival of Western agriculture."[16]

Today, the debate over agricultural sustainability and plant genetic resources is usually framed in terms of public vs. private innovation and control. Yet for the first generation of activists, the emergence of large public research centers, staffed with scientists overwhelmingly hailing from the Global North and funded by Northern governments and private foundations, was part of the larger picture. This included the global network of international agricultural research centers under the Consultative Group on International Agricultural Research (CGIAR), the creation of which, to quote Mooney again, "had been a blunt move [...] to wrest control of agricultural development from FAO and place it in the hands of a manageable scientific élite."[17] For Kloppenburg, they were mere "vehicles for the efficient extraction of plant genetic resources from the Third World and their transfer to the gene banks of Europe, North America, and Japan" akin to "the eighteenth- and nineteenth-century botanical gardens that served as conduits for the transmission of plant genetic information from the colonies to the imperial powers."[18]

The latter quote clearly demonstrates how criticisms of agricultural transformations were bound up within the larger discursive framework of anti-colonialism. Open access to plant genetic resources and both the knowledge systems and social contexts in which their utilization had traditionally taken place were considered to be under threat not only from intellectual property rights as such, but from the larger trend towards the commercialization and industrialization of agriculture on behalf of Northern governments and multinational companies. The "solution" developing countries began to pursue from the late 1980s on would shift the playing field by asserting sovereign control over genetic resources.

State sovereignty and access regulation

In the early 1990s, state sovereignty over genetic resources emerged as a principle of international law, first under an amendment to the International Undertaking.[19] In 1992, state sovereignty, and with it the right of provider countries to regulate third-party access to their genetic resources and require the fair and equitable sharing of the benefits arising out of their utilization, became a central pillar of the CBD, an international treaty originally intended exclusively as an instrument for the protection of biodiversity. Linking access to, and benefit-sharing from, genetic resources to biodiversity protection within a single instrument is not necessarily a foregone conclusion. Attempts to conserve biodiversity and use it in a sustainable way were triggered by the increasing recognition of man-made impacts on species and ecosystems worldwide, the "holocene extinction" as it became known in the 1980s. While different sectoral treaties had previously been concluded in this area, for instance for the trade in endangered species, migratory species or the protection of wetlands,[20] the CBD was to provide an integrated and comprehensive regime for global biodiversity as a whole. Yet there is no functional need for linking those questions to the utilization of genetic resources. Instead, the CBD was built around a compromise: industrialized countries wanted to protect biodiversity, particularly in the highly diverse countries of the Global South, whereas developing countries sought a solution to the access and utilization problem. As companies from industrialized countries were using the open access to genetic resources the International Undertaking had granted for obtaining raw materials for commercial products protected by intellectual property rights, the answer for many developing countries was simple: access needed to be regulated. Only then could they ensure participation in the (commercial) benefits arising from their utilization.

The CBD defines genetic resources as "any material of plant, animal, microbial or other origin containing functional units of heredity" that is "of actual or potential value."[21] This includes materials in their *physical* form and, potentially, the genetic *information* they carry, with genetic sequence data thus, depending on the legal interpretation, also falling within the CBD's scope.[22] The sovereignty of nation states over their genetic resources is independent of where the latter are located and applies equally to those materials (collected after the CBD's entry into force in 1993) stored in *ex situ* collections abroad. The coverage of "any material" of "plant, animal, microbial or other origin" means that the Convention has an extremely broad scope, yet in 1995, the CBD's COP decided that it does not cover human genetic resources.[23]

The CBD applies both to genetic resources under national jurisdiction and those beyond. However, its provisions on ABS relate exclusively to the former. Article 15 holds that "the authority to determine access to genetic resources rests with the national governments." Such access requires that an initial authorization is granted to a prospective user (Prior Informed Consent) and that a bilateral contract is drawn up specifying the terms of use and, importantly, the terms of benefit-sharing between user and provider country (Mutually Agreed Terms).[24] These provisions are the lever for attaining one of the Convention's three objectives, the fair and equitable sharing of the benefits arising out of the utilization of genetic resources.[25] Fairness and equity are not defined but create a procedural standard: benefit-sharing is fair and equitable when Prior Informed Consent has been granted and when Mutually Agreed Terms have been established and are being adhered to.[26] The CBD leaves open which and how many benefits are to be shared from the user to the provider country; providers and users are to agree under bilateral contracts. It may include, for instance, up-front payments, royalties or a share of the commercial profits generated from inventions incorporating the relevant genetic resources or non-monetary benefits such as the sharing of research results or the cross licensing of intellectual property rights to the provider country.

In order to balance the interest of provider countries in obtaining benefit-sharing streams with the interest of users in access, article 15.2 requires contracting parties "to create conditions to facilitate access to genetic resources for environmentally sound uses." Thus, while states enjoy sovereignty over their genetic resources, the CBD prevents the frivolous use of the Prior Informed Consent requirement to arbitrarily deny requests for access.

The reason for developing countries to pursue the benefit-sharing objective and the state sovereignty principle was their assumption that genetic resources have vast commercial value, which is not necessarily the case.[27] The concern to be passed over in the global biotechnology bonanza thus displaced earlier concerns over the growing number of intellectual property claims over plants, cells and other genetic resources. In fact, the CBD implicitly legitimizes intellectual property claims: commercial users only generate benefits, some of which can subsequently be shared, if they have the legal means to protect their inventions. The choice for accepting intellectual property rights in exchange for benefit-sharing, instead of attempting to limit their anti-commons effect, was a decision of monumental importance shaping subsequent regime formation in the issue area.

Biopiracy, misappropriation, misuse

"Biopiracy" is the key narrative driving developing countries' engagement with ABS governance. Yet it is an ambiguous concept. It is not defined under international law nor are its economic dimensions well understood. The concept can be understood in two different ways. On one hand, it can refer to the enclosure of the genetic commons, such as the "misappropriation of indigenous peoples knowledge and biocultural resources, especially through the use of intellectual property mechanism."[28] That is, biopiracy is the very act of claiming intellectual property rights over genetic resources. On the other hand, biopiracy can be understood as the appropriation of genetic resources "without recognition, reward or protection to informal innovators"[29] or "the unauthorized extraction of biological resources and/or associated traditional knowledge from developing countries, or to the patenting of spurious 'inventions' based on such knowledge or resources without compensation."[30]

Those two accounts of biopiracy, the claiming of intellectual property and the claiming of intellectual property without authorization and compensation, are mutually exclusive. They respectively correspond to the common heritage paradigm and the principle of access regulation by nation states (or local communities) sketched out above. The concept was originally not intended to have a precise meaning, though. Instead, it emerged as "a counter-concept to the intellectual property 'piracy' that the emerging economies were being condemned for" during the course of the 1990s.[31] In that sense, it is similar to the concept of Farmers' Rights, which I discuss in Chapter 5, which was intended as a counter-regime norm against the strengthening of plant breeders' rights since the 1970s and 1980s. The term is rooted within an anti-colonialist narrative, specifically in dependency theory.[32]

What are we to make of the concept? Here, I adopt a legal perspective on biopiracy, that is, access to and utilization of genetic resources in breach of national or international laws. Thus, I distinguish between the misappropriation and the misuse of genetic resources.[33] Those comprise different acts that also require different types of (legal and institutional) remedies. "Misappropriation" means access to, and utilization of, genetic resources without authorization, that is, without the user's obtaining Prior Informed Consent in those cases where providers require it. Misappropriation applies both to resources from within a provider country's territory and those held in *ex situ* collections abroad, which, nevertheless, are subject to the provider's sovereignty. "Misuse," conversely, is the utilization of genetic resources in breach of the Mutually Agreed Terms established between the user and the provider. This may be the failure to adequately share benefits in line with contractual arrangements or the utilization of genetic resources for unauthorized purposes.

Misappropriation and misuse are of a different legal nature. The former breaches the provider's public law; the latter violates contractual arrangements under private law. This has significant implications for enforcement: effectively addressing misappropriation requires that a provider country's domestic access legislation be enforced in the jurisdiction where utilization takes place. International commercial contracts, conversely, usually include provisions on the applicable law for resolving disputes between the parties to the contract. They are thus enforceable without the country in which utilization takes place taking any additional action.

This legal approach to biopiracy does not capture the concept's normative substance yet is a useful starting point when analyzing international negotiations on the compliance provisions of different ABS regimes. Yet this also requires taking into account the reasons users might misappropriate or misuse genetic resources in the first place. The concept of biopiracy implicitly assumes intentionality: users do not comply with their requirement to obtain Prior Informed Consent or establish Mutually Agreed Terms because they expect the costs of compliance to exceed the risk of non-compliance. In other words: monitoring and enforcement are insufficient for deterring users from misappropriation and/or misuse.

This is a one-sided way of looking at the problem. As industrialized countries and observers have frequently pointed out, one reason for misappropriation (if not misuse) might be the lack of clear, transparent and efficient access legislation in provider countries. Prospective users might be unsure which legal requirements they are, in fact, expected to comply with, or they might obtain genetic resources from *ex situ* collections being unaware of the necessity to obtain Prior Informed Consent or the knowledge of how to do so, possibly due to a lack of documentation regarding the whereabouts of a genetic resource and the applicable laws. Particularly in the negotiations on the Nagoya Protocol (Chapter 6), the streamlining of national access regulations was a major objective for user countries.

Those two ways of looking at the problem of biopiracy hark back to the academic debate on enforcing versus managing compliance.[34] The threat of punitive sanctions will not resolve misappropriation (or misuse) where it results from the inability to comply as opposed to the unwillingness to do so. Conversely, facilitation is insufficient where users base their decisions on the expected costs of compliance versus the expected costs of non-compliance. This issue is also related to the *amount* of benefits users are expected to share. For instance, where benefit-sharing agreements foresee the transfer of relatively limited amounts of monetary benefits from users to providers, non-compliance will be less of an issue than where significant financial stakes are involved.[35]

This brings us to the next question: what is the actual extent of biopiracy, and how large are the resulting opportunity costs, in terms of lost benefits, to providers? The answer is that nobody knows. There is neither a global tracking mechanism for monitoring the utilization and transfer of different types of genetic resources through the value chain, nor can non-compliance with national access regulations or benefit-sharing contracts readily be assessed. The illegality of such actions would also give users a reason to hide the fact as such. Discussions merely rely on anecdotal evidence. This includes a number of well-documented cases, such as the Hoodia plant, used as a thirst and appetite suppressant by the native San people of the Kalahari desert, which was misappropriated by industry for the purpose of creating a natural diet pill; a patent on a fungicide derived from extracts of the Indian Neem tree, an application long known to local communities; the attempt by a US company to patent a type of Basmati rice that had been grown in South Asia for centuries; and the case of the Enola bean, where a US businessman obtained a utility patent on the bean, its plant, pollen and method of production, subsequently demanding licensing fees from farmers wishing to export those beans to the US market.[36]

Those examples show the difficulties, and sometimes confusion, in the debate on biopiracy. With the exception of the Hoodia plant, the examples mentioned above are instances where patent applications were filed (and, in three cases, granted) despite a lack of novelty and inventive step. They do not exemplify problems with patent law as such, but rather problems of the erroneous *application* of patent law. At the same time, they are widely considered prime instances of biopiracy, which, depending on its specific definition, is intended to highlight problems with patent law *per se*. This is contradictory: had proper patent examination taken place, "biopiracy" would not have happened in the first place.

What can be said about those cases, instead, is that the users of genetic resources failed to obtain Prior Informed Consent and Mutually Agreed Terms. This is not a problem of patent law but one of the insufficient implementation of the CBD. Then again, the ineligibility of Basmati rice, the Enola bean and the Neem fungicide for patent protection is unrelated to the question of whether or not Prior Informed Consent has been granted and Mutually Agreed Terms established—even had the respective inventors done so, it would not have made their "inventions" eligible for patent protection.

Yet the four examples show a more fundamental ambiguity of the concept of biopiracy as denoting either the claiming of intellectual property or the failure to share the benefits arising out of the utilization of genetic resources, the Hoodia case being an instance of the latter. These two interpretations are mutually exclusive: if patents (or plant variety rights) are not taken out on inventions resulting from genetic resources, there will simply be no benefits to subsequently be shared. The commercial utilization of the Hoodia plant, first by the South African Council for Scientific and Industrial Research, later by Unilever, involved substantial investments. Without the option to protect their intellectual property, it is unlikely that those investments would have taken place.

This ambiguity in the relationship between access to genetic resources and intellectual property rights on the one hand and benefit-sharing and intellectual property rights on the other greatly complicates international ABS governance and has led to substantial inconsistencies within the first ABS regime I analyze in this book: the FAO Seed Treaty.

Notes

1 Rabitz (2015).
2 Muzaka (2010).
3 Muzaka (2011), 24–31.
4 Abbott and Dukes (2009), 3–4.
5 Ceccarelli (2009).

6 see Drahos (1996).
7 Lenk (2007), 124.
8 Shapiro (2001); Murray and Stern (2007).
9 Convention for the Protection of Cultural Property in the Event of Armed Conflict, Article 1.a; Treaty on Principles Governing the Activities of States in the Exploration and Use of Outer Space, Including the Moon and Other Celestial Bodies, Article 1 and 2; Antarctic Treaty, preamble; United Nations Convention on the Law of the Sea, Article 137.
10 FAO Conference, Resolution 8//83, Annex, Article 1.
11 International Undertaking, article 2.1(a).
12 Andersen (2008), 91.
13 Diamond (2005); Murphy (2007).
14 Kloppenburg (1987). The term 'seed wars' was coined by a journalist writing for the Financial Times as an ironic take on Reagan's 'star wars' missile defense program.
15 Shiva (1993), 63.
16 Mooney (1983), 10–11.
17 Mooney (1983), 66.
18 Kloppenburg (2004), 161.
19 Resolution 3/91.
20 see Gomar et al. (2014) for an overview.
21 CBD Article 2.
22 Tvedt and Schei (2014).
23 Decision II/11, para 4. While the COP technically did not 'decide' but 'reaffirmed' the exclusion, neither did previous decisions nor the Convention's text 'affirm' it in the first place.
24 CBD articles 15.1, 15.4 and 15.5.
25 CBD Articles 15.1 and 1.
26 Parties may also waive their right to regulate access and require benefit-sharing.
27 ten Kate and Laird (2000).
28 Mgbeoji (2006), 12.
29 RAFI (2001), 72.
30 Dutfield (2004), 52; see also Shiva (2001) and Robinson (2010).
31 Dutfield (2014), 650.
32 i.e., Frank (1978); Wallerstein (1983).
33 Rabitz (2015).
34 Chayes and Chayes (1993); Downs et al. (1996).
35 Rabitz (2015).
36 See Mgbeoji (2006); Wynberg et al. (2009): Robinson (2010).

5 Plant genetic resources for food and agriculture

The first case study of this book deals with the processes leading from the 1983 International Undertaking on Plant Genetic Resources ("International Undertaking") to the 2001 Seed Treaty. The Undertaking, a legally non-binding resolution adopted by the FAO Conference, was intended as an open-access regime under which Plant Genetic Resources for Food and Agriculture (PGRFA) were to be available without restrictions and for the benefit of humanity writ large. Yet the key problem of the Undertaking was: should open access to PGRFA apply exlusively to the genetic raw materials used for the breeding of new plant varieties, or should improved varieties, brought under intellectual property protection, equally be available without restrictions?

The Undertaking went through a series of transformations between 1989 and 1991, as parties adopted amendments on plant variety rights, Farmers' Rights and the sovereignty of nation states over their PGRFA. In 1992, the adoption of the CBD codified the principle of state sovereignty over genetic resources, including PGRFA, for the first time in a legally binding international treaty.[1] In 1994, negotiations commenced within the FAO on revising the International Undertaking to bring it in line with the CBD; this process culminated with the adoption of the Seed Treaty in 2001.

The Seed Treaty created a global network of seed banks from which certain PGRFA are available for breeding purposes. While acknowledging sovereign rights over PGRFA, access takes place under the (non-negotiable) terms of a Standard Material Transfer Agreement (SMTA). Where users improve such PGRFA and commercialize them under patent protection, they are obliged to share parts of the accruing profits on a multilateral basis. Where the International Undertaking was not more than a collection of unspecific principles within a non-binding recommendation later amended in a haphazard and inconsistent way, the Seed Treaty created the first operational and private contract-based ABS regime at the global scale.

The processes leading from the International Undertaking to the Seed Treaty played out within the wider context of the transformation of global agriculture, which had begun in the 1950s and, with the emergence of commercial biotechnology and changes in intellectual property law, intensified during the 1980s. Multiple, interrelated problems defined the conflict lines not only between the Global North and the Global South, but also between exporters and importers of agricultural products and between countries rich in agricultural biodiversity and those particularly dependent on PGRFA from abroad. First, the dramatic decline in agricultural biodiversity, at the global scale, prompted collective efforts for the better conservation and sustainable use of PGRFA. Second, the increasing dominance of large agricultural companies hailing from the Global North, together with the strengthening and international diffusion of regimes for plant variety rights, led to attempts to protect the autonomy and ways of life of small-scale farmers and agricultural communities in developing countries. Third, in the way that agricultural innovation in the private sector began to take off, the public sector entered into long-term and worldwide decline. The breeding of new plant varieties was thus increasingly driven by commercial motives. Fourth, the promises of large commercial profits from agricultural biodiversity gave rise to conflicts between providers of PGRFA hoping to cash in on the utilization of "their" genetic raw materials abroad, whereas the governments of countries with strong biotechnology sectors sought to maintain unrestricted access to PGRFA on behalf of their domestic users.[2]

As the case of the Seed Treaty is often framed in terms of a North-South divide, it is important to note from the outset that, from a situation-structural perspective, the objectives of large providers (such as Brazil and China) were defensive in nature and did not align with that of many other developing countries. In terms of the distribution of interests, the only genuinely "reformist" actors were those with neither significant *ex situ* holdings of PGRFA nor capacities for their commercial utilization. In terms of interdependence, the general pattern within which all countries depend on each other for access to plant germplasm co-exists with a more specific one in which the cooperation of large provider countries is necessary in order to obtain access to their PGRFA holdings, and the cooperation of large user countries is necessary for the realization of benefit-sharing flows and potential limitations to intellectual property claims on seeds and their genetic parts and components.

Agriculture, genetic erosion and plant breeding

The problem of access to PGRFA and the sharing of resulting benefits forms part of the fundamental transformation of global agriculture that has been playing out since the Green Revolution of the first half of the 20th century. In parts, this is the history of science and technology. Advances in chemistry allowed for the synthesis of nitrogen fertilizers on an industrial scale since the beginning of the century. Organochlorine pesticides such as the infamous DDT came to be in widespread use after the second World War. Since the 1920s, high-yield hybrid crops created by crossing inbred lines of different species quickly came to dominate corn production in the US and subsequently in other parts of the world; similarly, radiation and the use of synthetic chemicals allowed for inducing high levels of genetic mutation in plants, rendering large numbers of useful traits in a short period of time; and artificial chromosome doubling enabled the creation of fertile varieties of (otherwise sterile) crosses.

Advances in plant breeding and agricultural chemistry were instrumental in overcoming food crises, which was predicted by accounts such as Paul Ehrlich's *Population Bomb* or the Club of Rome's 1972 report on *The Limits to Growth*. The most spectacular outcome of the Green Revolution, the dwarf wheat developed by a multinational team of scientists working under the American biologist Norman Borlaugh at the International Maize and Wheat Improvement Center in Mexico, made India and Pakistan, which were facing severe food shortages during the 1960s, into net agricultural exporters within the course of a few years.

Yet the industrialization of agriculture came at a price: the loss of agricultural biodiversity, or "genetic erosion," is a direct consequence of the shift towards monocultural agricultural systems intended to achieve economies of scale. Such genetically uniform plantations are not only more vulnerable to pests and soil erosion, but they provide only limited genetic raw materials for the production of new plant varieties. The lower the genetic diversity of input crops, the less likely is it that new and useful traits can be produced. As countries are highly interdependent on each other for obtaining input materials and improving the plant varieties they use, genetic erosion poses a global risk.[3]

Yet access to genetic inputs is at risk not only from genetic erosion. The changes in intellectual property law sketched out in Chapter 3 also limit access to plant germplasm insofar as authorization by the right holder is required. Both plant patents and plant variety rights have been the subject of enormous controversy since the 1980s. Its proponents argue that only the prospects of a limited period of exclusive rights over new plants or plant varieties create the proper incentives for profit-oriented actors to undertake the quite significant financial risks involved in contemporary plant breeding.

The critics point out that the free exchange of plant germplasm has worked reasonably well for the last 12,000 years or so[4] and that the privatization of genetic materials developed and cultivated by farming communities all over the world and for significant amounts of time raises important questions of equity.

Intellectual property thus became the focal point in the seed wars of the 1980s, when developing countries and civil society organizations voiced concern that "the high-yield varieties and the fertilizers, herbicides and pesticides necessary for their cultivation had to be bought for money from private companies of the North" whereas PGRFA "flowed gratuitously into the North."[5] Developing countries and a handful of activists thus began pushing the notion of Farmers' Rights to reflect that "[j]ust as plant scientists are entitled to a reward for their labour in creating breeding lines and elite varieties, so farmers have a right to a reward for creating and maintaining land races and other 'raw' plant genetic resource."[6]

Criticisms of intellectual property rights pertained to both patents and UPOV-style plant variety rights, although the former has drawn significantly more ire from activists, intellectuals, researchers and civil society movements. In fact, the broad international consent that patents are an inadequate instrument for agricultural innovation led to the creation of UPOV in the first place. As was discussed in Chapter 3, the narrowing of the breeders' exemption under the 1991 version of UPOV did bring it closer in line with generic utility patents. At the same time, though, the terrain of plant variety rights is increasingly being encroached upon by the patent system. As Chapter 3 has shown, the domain of plant variety rights has been continuously shrinking in Europe by expanding the scope of patentable subject matter. In the US, plant variety rights play a fairly insignificant role due to the full availability of patents. However, the precise impact of intellectual property rights on agricultural systems remains largely unclear. The debate on plant patents and plant variety rights, however, frequently intermingles both economic and ethical factors while lumping together intellectual property rights as a matter of public policy with more fundamental, and equally normatively charged, criticisms of globalization and market liberalization.

The two dimensions, genetic erosion and access restrictions from intellectual property rights, are linked, with an increasingly small number of agricultural MNCs supplying global markets with a limited number of blockbuster seeds under intellectual property protection (plant varieties or, where available, patents). The adoption of the CBD in 1992 further complicated this picture. State sovereignty over genetic resources and the authority of provider countries to regulate access implied that, all of a sudden, additional legal barriers might stand in the way of access to breeding materials. This raised the problem that, if provider countries imposed tight regulations in order to obtain large benefit-sharing flows from users of their PGRFA, this would further undermine agricultural innovation. While the state sovereignty approach had been a countermeasure against the increasing privatization of genetic resources through intellectual property rights, in the case of PGRFA, the paradigm threatened to further undermine the breeding of new plant varieties. State sovereignty thus raises a classic collective action problem: while all countries are dependent on the exchange of plant germplasm, individual providers have incentives to foreclose access to their PGRFA in order to maximize their respective pay-offs.

The at times complex legal debate on intellectual property rights and state sovereignty should not detract from the core problem: genetic erosion and access restrictions through intellectual property rights and state sovereignty may impede agricultural innovation at precisely the time when projections for world population growth are regularly being corrected upwards; changing consumption patterns in emerging economies lead to growing amounts of agricultural produce being used as feed; the increasing incidence of droughts and flooding as well as water and temperature stress as a consequence of climate change is likely to harm global food production; and biofuel production is raising concerns on the conversion of increasingly sparse arable land.[7] The timing could thus not be worse for protracted international discussions on who gets to use genetic raw materials for crop improvement and under which conditions.

Situation structure

The situation structure in PGRFA is defined by countries' relative capacities to provide genetic input materials for plant breeding, to utilize them for the breeding of new varieties and bring them under intellectual property protection.

To start with the first point, countries with a large diversity of PGRFA of high commercial value (i.e., "cash crops" such as oranges, potatoes or coffee) or particular importance to global food security enjoy a privileged position within the situation structure as others depend on having access to those resources, which may be held *in situ* or *ex situ* in seed banks. Yet to maximize their individual gains from international cooperation in ABS, provider countries prefer bilateral, rather than multilateral, benefit-sharing.

While the precise extent and distribution of *in situ* and on-farm agricultural biodiversity is difficult to assess and few countries have created comprehensive inventories, centers of diversity for major crops are Africa (Cassava and Sorghum), South and Central America (Bean, Cassava, Maize, Potatoes and Tomatoes), South Asia (Cassava, Rice, Wheat), East Asia (Rice and Wheat) and Western Europe (Beet and Rapeseed).[8] In many regions of the world, the state of *in situ* conservation is poor, yet local communities continue to conserve and manage significant amounts of crop genetic diversity on-farm. Major challenges to *in situ* and on-farm conservation include climate change, habitat destruction, invasive alien species, the displacement of traditional and diverse varieties by modern and genetically uniform ones and possibly transgene flow from genetically modified crops.[9]

Ex situ PGRFA are dispersed across a wide range of seed banks in numerous countries. Storage is technically demanding. Cryopreservation enables some seeds to be stored over extremely long periods of time: up to hundreds of years or even millennia. Especially for tropical crops, however, DNA degradation makes repeated replanting necessary.[10] Seed banks require elaborate electronic catalogs for ensuring that PGRFA can be efficiently accessed; financial support and highly trained scientific staff, conditions that are frequently not met in developing countries.[11] Guinea recently lost its entire *ex situ* collections due to electricity shortages.[12]

The largest and best-equipped *ex situ* collections under government control are found in China (~351000 accessions from 735 species, holding 56% of global accessions of millet and 14% of soybeans); the US (~536000 accessions from 13,000 species, accounting for ~7% of global germplasm holdings); India (~349000 accessions from 1187 species) and Russia (~380000 accessions from 2500 species). Yet nation states are not the only, or even the largest, providers of PGRFA. The biggest collections are presently managed by the Seed Treaty's Article 15 institutions: the research centers of the CGIAR and similar centers. The multinational Millennium Seed Bank Partnership, located in the UK, holds close to two billion seeds from 35,000 different species. The Svalbard Global Seed Vault, located on the island of Spitsbergen, holds more than 8000 samples from 4000 species, which are duplicates from collections all over the world as an insurance policy against the failure of other seed banks.

While no historical data is available, presently more than half of the PGRFA available under the Seed Treaty are held by the CGIAR centers and roughly one-third are stored in public seed banks in the North American and European region. A marginal part is also being provided from private collections (see figure 5.1).[13]

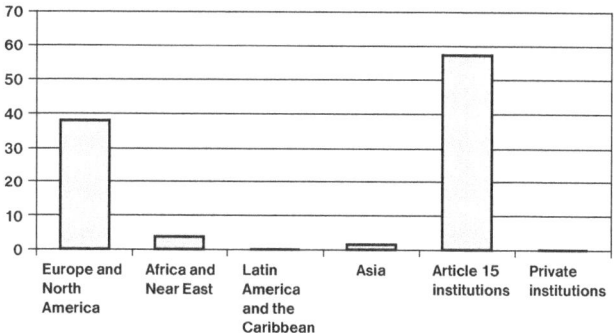

Figure 5.1 Annex I PGRFA holdings by provider
Source: FAO.

Simply looking at the absolute numbers of accessions can be misleading. PGRFA differ in their commercial and nutritional relevance, making some more important than others. For illustration, Table 5.1 lists the world's top 5 holders of selected staple crops as of 2010. The largest holders of those crops are the CGIAR institutions and a handful of nation states.

With some idiosyncratic exceptions, Brazil, China and India are the only developing countries holding significant amounts of valuable staple crops. In recent years, these three have slowly been building up a sizable domestic agrobiotech industry.[14] Yet for the period from 1983 until 2001, the focus of this chapter, they constitute the largest providers of PGRFA that are not simultaneously major users.

Commercial value from the utilization of PGRFA is chiefly generated in countries with sufficient biotechnological and breeding capacities. Looking at market structure, the agribusiness sector has been marked by an extraordinary amount of corporate consolidation in recent decades. In 2011, the 10 largest players in global agribusiness, jointly accounting for 75.3% of global seed sales, were headquartered in merely 5 countries: France, Germany, Japan, Switzerland and the US.[15] Market concentration is bound to increase further with the planned takeovers of Dupont by Dow Chemicals, Syngenta by ChemChina and Monsanto by Bayer. Relative capacities for utilizing PGRFA can also be measured in terms of the ownership of relevant intellectual property. Unfortunately, no global statistics exist for plant variety rights, which is mainly due to differences in national seed laws. Yet data exists for breeding technologies. A search in the Patentscope® database for the number of patents on genetic-engineering technologies for plants by applicants' country of residence in the period from 1994 (when negotiations on the Seed Treaty commenced) to 2001 (when they concluded) shows that the US alone accounts for almost half of all relevant patent publications. The highest-ranked non-OECD country, India, accounts for a mere 0.005%.[16] Those numbers do not cover traditional (non-biotechnological) breeding methods, yet they show that ownership of the technologies that came out of the biotechnological revolution of the 1980s is distributed extremely unevenly (figure 5.2).

Table 5.1 Top 5 holders of selected cash crops

Crop	Rank 1	Global share	Rank 2	Global share	Rank 3	Global share	Rank 4	Global share	Rank 5	Global share
Wheat	CGIAR	13%	USA	7%	China	5%	India	4%	CGIAR	4%
Rice	CGIAR	14%	India	11%	China	9%	Japan	6%	Korea	3%
Barley	Canada	9%	USA	6%	Brazil	6%	CGIAR	6%	Japan	5%
Maize	CGIAR	8%	Portugal	7%	USA	6%	China	6%	Mexico	4%
Beans	CGIAR	14%	USA	6%	Brazil	88%	Mexico	5%	Germany	3%
Sorghum	CGIAR	16%	USA	15%	China	8%	India	7%	Ethiopia	3%
Soybeans	China	14%	USA	9%	South Korea	8%	CGIAR	7%	Brazil	5%
Oats	Canada	21%	USA	16%	Russia	9%	Germany	4%	Kenya	3%

Source: FAO (2010), 62–64.

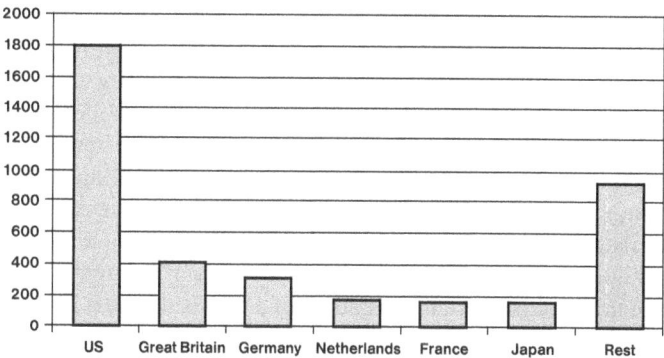

Figure 5.2 Patent publications on plant breeding technologies by applicants' country of residence, 1994–2001
Source: Patentscope(r).

The situation structure in PGRFA does not neatly divide North from South. Instead, countries cluster into three groups: countries that provide PGRFA yet had low capacities for their commercial utilization in the period under consideration ("providers" such as Brazil, China and India); countries with large and technologically advanced *ex situ* collections as well as capacities for commercial utilization ("users/providers" such as the USA, Japan and some EU member states) and countries with neither significant *ex situ* collections nor capacities for commercial use ("non-users/non-providers" such as many African and Pacific countries). In line with the situation-structural approach, this allows for deducing both interests and patterns of interdependence as follows:

Providers have an interest in their PGRFA being commercially utilized abroad and maximizing the benefits that flow back to them. This implies acceptance of intellectual property rights as a pre-condition for the generation of commercial value. It also implies a preference for bilateral over multilateral benefit-sharing, secured with measures for monitoring and compliance. However, the spatial distribution of PGRFA across multiple countries and *ex situ* collections, some of which are under the control of foreign governments, others under the auspices of international organizations such as the CGIAR, reduces their leverage as users may be able to source samples elsewhere. In addition, monitoring the transnational flow and utilization of PGRFA is impossible without the cooperation of the countries where utilization takes place. As the CBD, in principle, already provides for such a regime, this group of actors has the strongest status-quo orientation: while access to PGRFA in other collections, such as those of the CGIAR, is clearly important, the purpose of the PGRFA regime is to maintain the scope of the CBD as widely as possible and create a clear delineation between the two.

Users/providers generate commercial value from PGRFA and are thus interested in minimizing constraints on access through monitoring and compliance measures; ensuring that intellectual property rights are fully available and minimizing benefit-sharing obligations.[17] For money to flow through the system, the cooperation of those actors is indispensable. While users/providers depend on access to PGRFA, as noted above, alternative sources may be used if individual providers decide to regulate access to their *ex situ* collections.

Non-users/non-providers have an interest in limiting the negative externalities from intellectual property rights when seeds used domestically are protected or when their agricultural products face entry barriers due to intellectual property protection in export markets.[18] Similarly, access regulations in line with the CBD threaten to limit the free exchange of germplasm on which domestic farmers depend.[19] Large gains from bilateral benefit-sharing arrangements are unavailable and multilateral benefit-sharing reduces the pay-offs each individual country receives. Without the ability to provide large amounts of raw materials or share benefits from improved materials, this group of actors is the most insignificant in terms of situation structure.

Negotiations

This section commences with the 1983 International Undertaking and its amendments in 1989 and 1991 before turning to the negotiations on what would later become the Seed Treaty. Specifically, I address Farmers' Rights, the 1999 Montreux package deal, the linkages between access and intellectual property rights, as well as between benefit-sharing and intellectual property rights; the negotiation of the regime's scope and the adoption of the Treaty in 2001.

Amending the International Undertaking

Adopted by an FAO Conference resolution in 1983, the International Undertaking took an open-access approach to plant genetic resources as the common heritage of mankind; being available without restrictions, their utilization was to benefit both present and future generations. The Undertaking was a reaction by the governments of the Global South who were concerned about the South-North flow of genetic resources and their subsequent transformation into private property.[20] Yet major industrialized countries, including Germany, France, the UK and the US, did not adhere to the Undertaking, or did so only with reservations, due to the incompatibility of open access to plant genetic resources with the proprietary rights granted by plant variety protection.[21] The reason was that the Undertaking's scope was not limited to "raw" genetic resources but also covered "special genetic stocks" such as elite breeders' lines. This included, for instance, the in-bred lines used in the production of hybrid seeds—high-yield seeds that lose their vigor in the second and subsequent generations, thus requiring farmers to regularly purchase new stocks from agricultural suppliers, a business model that requires the in-bred lines for producing hybrids be proprietary. The plant breeding industry and several industrialized countries thus considered the Undertaking incompatible with plant breeders' rights in particular and private property in general.[22] Accordingly, several industrialized countries, including Canada, France, Japan, the UK and the US, chose not to, or to only partially, adhere to the Undertaking.

Yet without the biotechnological powerhouses of the Global North making their improved genetic resources available, the Undertaking would ultimately fail to achieve its objective of unrestricted access to improved plant varieties. In order to broaden participation, Resolution 4/89 acknowledged the compatibility of UPOV-style plant variety rights with the Undertaking; at the same time, resolution 5/89 acknowledged Farmers' Rights as "rights arising from the past, present and future contributions of farmers in conserving, improving, and making available plant genetic resources [...] as trustee for present and future generations of farmers."[23] Developing countries fiercely opposed to intellectual property rights on seeds thus "gained recognition of Farmers' Rights in exchange for something that already existed,"[24] in the hopes of facilitating the participation of those countries concerned about "possible conflict [...] with their international obligations and existing national regulations."[25]

The two resolutions contained embryonic rules on access and benefit-sharing. First, resolution 4/89 emphasized that "free access" to plant genetic resources would not necessarily imply that access is "free of charge."[26] On one hand, the provision can be regarded as a concession to plant breeders, that is, third parties would be charged for access to plant varieties under intellectual property protection, as is the normal practice with licensing agreements.[27] On the other, charging for access also entails the possibility of providers of genetic raw materials requiring some form of monetary arrangement. The resolution also held that benefits derived from plant genetic resources should be part of a "reciprocal system" and limited to participating countries.[28] Resolution 5/89, in turn, decided that farmers, farming communities and countries should "participate fully" in the benefits arising from the utilization of plant genetic resources in plant breeding and biotechnology more broadly.[29]

These initial rules on access and benefit-sharing were complemented by resolution 3/91, which introduced the principle of state sovereignty over plant genetic resources into the Undertaking while at the same time holding on to the common-heritage principle. Squaring common heritage with state sovereignty is a tricky exercise. While common heritage can be interpreted as requiring states, in exercising their sovereignty, to act to the wider benefit of the international community or humanity writ large,[30] sovereignty also entails the right of provider countries to regulate, and thus ultimately to deny, access to plant genetic resources. Thus, "[t]he Undertaking's original goal of 'unrestricted access' has been steadily narrowed over time."[31]

The amendments to the Undertaking show, first, the role of issue linkages in ensuring the participation of those actors able to make critical contributions in a given area of international cooperation. Including actors with diverging interests thus increased the institutional scope by creating a direct interface with plant variety rights. Second, the amendments are a clear instance of institutional layering, as reformist actors added novel sets of rules to the pre-existing institutional core. Third, this accords with the idea that new institutional layers can have a transformative effect on the original institution. On one hand, the Undertaking in its 1991 version is closer to the approach taken by the CBD than its original

(1983) version. On the other hand, institutional layering can also lead to inconsistencies, such as in the problematic relationship between common heritage and state sovereignty. As I will argue below, those inconsistencies are an important case in point against explaining institutional change in regime complexes in terms of interplay management.

Revising the International Undertaking

The conclusion of the CBD in 1992 raised the problem that bilateral negotiations for access to PGRFA would be incompatible with agricultural breeding practices. Thus, the 1992 Nairobi Final Act for the adoption of the CBD accordingly recognized "the need to seek solutions to outstanding matters concerning plant genetic resources."[32] Responding in turn, the FAO Council requested the FAO's Commission on Plant Genetic Resources (CPGR), in 1993, to provide a forum for harmonizing the International Undertaking with the CBD and considering "the issue of access on mutually agreed terms" to PGRFA as well as Farmers' Rights.[33] In 1995, CBD COP 2 recognized the "special nature of agricultural biodiversity," which would require "distinctive solutions," while declaring its support for the revision of the International Undertaking within the context of CPGR.[34]

One ordinary and two extraordinary sessions of the CPGR (since 1995 the Commission on Plant Genetic Resources for Food and Agriculture, CGRFA) in 1994 and 1996 set out to resolve those questions. Initially, the process was to consolidate the International Undertaking by formally incorporating resolutions 4/89, 5/89 and 3/91, which had previously been mere annexes to the main text.[35] In the second stage, the CGRFA would address access to PGRFA on mutually agreed terms, including to those resources collected before the CBD's entry into force, as well as Farmers' Rights. The final outcome of the process would be a non-binding agreement, akin to the International Undertaking; a freestanding, legally binding agreement or a protocol to the CBD.[36]

The central problem was how to design an international regime that is adapted to the specific nature of agricultural breeding yet legally consistent with the CBD. Talk of "harmonization" is deceptive. The amended Undertaking already contained key elements of the CBD, namely, state sovereignty and "reciprocal" benefit-sharing, yet the inconsistencies *within* the Undertaking implied that it was malleable and any revision process open-ended. Instead of harmonization, different groups of actors would henceforth attempt to institutionalize their respective preferences by pushing the process in one of several possible directions.

For countries such as Brazil, the Undertaking constituted "an effective loophole in the Convention on Biological Diversity"[37] by offering access to plant genetic resources without operational components for benefit-sharing. An overhauled Undertaking with a clear focus on a limited set of crops critical for food security would leave valuable materials within the scope of the CBD and ensure larger individual gains from bilateral benefit-sharing.[38] Countries with strong dependencies on foreign germplasm preferred a broad scope in order to enjoy

access to a larger range of materials on conditions that were better than those provided for under the CBD. The African Group, for instance, initially preferred the
revised Undertaking to grant broad coverage, limit intellectual property claims
and grant unconditional access to small-scale farmers.[39] Finally, industrialized
countries with strong agricultural industries, such as Australia, France and the
US, equally preferred broad coverage with intellectual property fully claimable
and benefit-sharing voluntary.[40] The key controversies that emerged were thus
related to scope (which PGRFA would fall under the new regime?); modalities
of access (which conditions, if any, should apply to recipients of PGRFA?); modalities of benefit-sharing (which types of benefits should be covered, and should
sharing be voluntary or mandatory?) and intellectual property (to what extent
would recipients of PGRFA be limited in their right to claim patents, plant variety
rights or both?).[41]

Farmers' Rights

Farmers' Rights had been on the agenda of developing countries and civil society organizations since the early 1980s and had first been acknowledged under
resolution 5/89.[42] It was also the first element of the Seed Treaty upon which
parties agreed. From 1995 onward, delegates in the CGRFA struggled with the
precise ways in which Farmers' Rights would be included in the Treaty, given the
concept's ambiguity.[43] Australia, Canada, the US and Japan preferred a minimalist approach under which national governments would merely support farmers
in the conservation and sustainable use of PGRFA. The EU took a more con-
ciliatory stance by granting the possibility that Farmers directly participate in
benefit-sharing under the later Seed Treaty. The most-comprehensive proposal,
put forth by developing countries, was the recognition of Farmers' Rights at the
international level (i.e., akin to a human right); international financial support for
the conservation, sustainable use and development of PGRFA; the protection and
promotion of collective rights in regards to farmers' "innovations, knowledge
and culturally diverse systems"; capacity building at the local level; participation
in relevant national and international decision-making processes and the review,
assessment and possibly modification of "intellectual property rights systems,
land tenure, and seed laws."[44]

Despite these strongly diverging positions, agreement proved surprisingly
easy. The draft article on Farmers' Rights coming out of the 1998 fifth extraordinary session of the CGRFA was heavily bracketed, with dissent ranging from
national versus international responsibilities over the bindingness of obligations
up to the question of whether the phrase "Farmers' Rights" should be capital-
ized or not.[45] Yet within the course of mere months, delegates had adopted a
draft article that would find its way into the final Seed Treaty almost verbatim.[46]
The later article 9 recognizes the contribution of farmers to global food security and assigns the responsibility for implementing Farmers' Rights exclusively
to national governments, which "should" protect Traditional Knowledge associated with PGRFA and allow farmers to participate in both benefit-sharing and

national decision-making processes relevant for the conservation and sustainable use of PGRFA.

The consensus reached between late 1998 and early 1999, even before the contours of the later Seed Treaty would emerge out of the 1999 Montreux meeting,[47] rids the concept of Farmers' Rights of large parts of its normative content. Both the timing and the nature of the concessions made by developing countries made have important implications for explaining institutional changes from the International Undertaking to the Seed Treaty, a point to which I will return below.

The Chairman's Elements

The contours of the Seed Treaty took shape in 1999 when an informal meeting in Montreux brought "the crucial breakthrough in establishing negotiating momentum."[48] The meeting resulted in the so-called Chairman's Elements, which formed the basis of the later Seed Treaty and its Multilateral System: the regime would grant access to a list of crops selected on the basis of food security and agricultural interdependence as well as the collections within the CGIAR network; PGRFA used for non-agricultural purposes, such as pharmaceutical research, would continue to fall under the CBD; access to PGRFA would not be unrestricted but "facilitated" without incorporating a tracking mechanism, and benefits would be shared multilaterally, primarily to developing countries and could include both non-monetary (technology transfer, capacity building, information exchange) and monetary components.[49]

The Chairman's Elements were adopted by the CGRFA in April 1999 and formed the basis for all subsequent negotiations. The Elements managed to resolve key legal problems and political conflicts. First, they carved out the later Seed Treaty's scope in relation to the CBD. The list-based approach and the exclusion of non-agricultural uses of PGRFA from the scope of the instrument meant that the scope of bilateral benefit-sharing under the CBD would remain wide so that providers of valuable genetic resources would be able to negotiate more-profitable bilateral agreements with users. The inclusion of the CGIAR collections offered the prospect of resolving the question of ownership, which had arisen since the Undertaking's resolution 3/91 and the conclusion of the CBD. Second, the choice for "facilitated" access in principle allowed for conditions to be imposed on recipients by obliging them to share benefits or by limiting their ability to claim intellectual property; it is also consistent with the obligation, under article 15.2 CBD, that parties "facilitate" access to genetic resources for environmentally sound uses. Third, the choice against a tracking mechanism prejudged any compliance mechanism the regime might come to include—parallel to the FAO process, discussions on the monitoring of the transnational utilization of genetic resources began sparking up in various other international forums starting in 1999. Finally, the acknowledgment that financial transfers would form part of the overall benefit-sharing mechanism under the Seed Treaty was an important concession towards developing countries.

As I discuss below, the Chairman's Elements are thus an important data point in favor of the interplay management hypothesis: while building on pre-existing commitments, they promised to resolve interface problems among the Seed Treaty, the CBD and the CGIAR. The same cannot be said, however, about the negotiations on the Seed Treaty's intellectual property components, which quickly flared up after Montreux.

Linkages with intellectual property rights

The Montreux meeting gave the negotiation process a definite direction. In April 1999 and on the basis of the Chairman's Elements, the CGRFA mandated a contact group to produce a negotiated text for adoption by the FAO Conference, with a total of six meetings taking place over the following two years. The contact group focused on two distinct issues: scope (see below) and the relationship between intellectual property rights and benefit-sharing, as well as intellectual property rights and access to PGRFA. These are two different types of linkages: on one hand, there is the age-old question of whether or not users accessing PGRFA from the Multilateral System would be restricted in their ability to claim patents or plant variety rights. On the other, is the question of whether intellectual property claims would lead to specific obligations to share the benefits out of the utilization of such PGRFA.

The linkage between intellectual property rights and benefit-sharing was easily resolved. Over the course of three negotiating sessions, the contact group found common ground on the Treaty's later Article 13.2(d)(ii): where PGRFA received from the Multilateral System and subsequently improved (i.e., through traditional breeding methods or biotechnology) are commercialized and third parties do not enjoy facilitated access to the materials due to intellectual property protection, the sharing of monetary benefits would become mandatory.[50] The idea originated from a proposal by the International Seed Federation,[51] which represents the world's largest commercial plant breeders and has been termed "a door-opener for intellectual property rights."[52] From the side of developing countries, it thus involved trading-off restrictions on intellectual property claims for the prospect of monetary benefits.

The question of how to design the linkage between intellectual property rights and access, however, haunted the negotiation process until the endgame. Members of the contact group disputed whether or not users accessing PGRFA from the Multilateral System should be prevented from claiming patents, or even plant variety rights, in the first place; whether such a prohibition would include the "parts and components," including isolated genes, of such PGRFA and whether it would apply to materials "received" from the Multilateral System or merely materials "in the form" received.[53] The language of the eventual article 12.3(d) was the direct cause of the US not acceding to the Seed Treaty. It also poses a theoretical puzzle, which neither interplay management, regime shifting nor situation structure can explain adequately. I return to this point below.

Notably, the linkages between access and intellectual property, on one hand, and benefit-sharing and intellectual property, on the other, are incompatible with each other. If benefits must be shared when certain types of intellectual property are claimed, this presupposes that such claims are permissible in the first place. Moreover, where commercial breeders are concerned, uncertain, long-term capital investments into plant breeding will not take place without the prospect of obtaining temporary monopoly rights on the eventual products. Limiting intellectual property claims thus limits the amount of benefits to be shared through the system. While the Chairman's Elements resolved the relationship of the future Seed Treaty with the CBD and the open-access approach taken by the CGIAR and the International Undertaking, the debate on intellectual property rights created yet another debate on the relationship between different property regimes, this time between private property and the state sovereignty paradigm.[54] The Seed Treaty would end up at the interface of all three: state sovereignty, private property and PGRFA as the "common concern of all countries"[55] and a narrowed conceptualization of the common heritage principle.

Annex I

In Montreux, the question of whether the Multilateral System would apply to *all* PGRFA or merely to a subset had been settled in favor of the latter. As Brazil put it at the second contact group meeting, the "window" the Seed Treaty would open up within the CBD regime should be "small" and with "clearly defined limits."[56] So, which PGRFA, precisely, should be included? Delegates faced a classic collective action problem in which all countries shared a common interest in contributing to food security by having suitable genetic input materials available, yet individually, each country had incentives to leave its most-valuable materials outside of the Multilateral System and within the ambit of bilateral benefit-sharing under the CBD.

Accordingly, negotiators took "a rather aggressive approach to adding or removing certain crops according to perceived national interest."[57] Two observers noted that, "China, the center of diversity for soybeans, insisted that soybeans be excluded, and when this was done, Latin America withdrew groundnuts. Not to be outdone, Africa took tropical forages off the table. This process may help the reader understand the irony of how a list of crops crucial to world food security contains asparagus and strawberries, but is missing soybeans, groundnuts, tropical forages and most 'poor people's' crops."[58]

While the inclusion of the CGIAR collections had been agreed on under the Chairman's Elements, a question negotiators had to resolve was how, or whether, to differentiate between listed and non-listed PGRFA held by the CGIAR centers: would their *entire* collections be available through the Multilateral System, or only those materials listed in the later Annex I? And, in case of the former, should access to non-Annex I PGRFA under the trusteeship of the centers be subject to the same conditions as in the case of listed materials?[59] At the same time that major provider countries hesitated to make their own PGRFA available in order to limit the size of the "window" in the CBD regime, they pushed for the mandatory inclusion of *all* (both listed and non-listed) CGIAR accessions in the Multilateral System.[60] As the CGIAR centers had previously granted access to their collections without requiring the sharing of benefits, this move promised to increase the economic value of the Multilateral System.[61]

As the CGIAR centers have their own legal personalities, the Treaty's article 15 "calls upon" them to conclude agreements with the FAO in order to bring their collections into the purview of the Treaty. Those agreements were concluded in 2006, after the Treaty's entry into force. Notably, both parties and non-parties to the Seed Treaty are able to access those collections on the same terms. That is, while those PGRFA provided by contracting parties are only accessible by *other* contracting parties and users within their respective jurisdictions, there is no such differentiation regarding the CGIAR collections, which, after all, were held in-trust for the international community at large, not merely those countries that decided to sign the Seed Treaty. Pursuant to articles 15.1(a) and (b), the CGIAR centers grant facilitated access to *all* of their PGRFA (Annex I and non-Annex I) collected before the conclusion of the CGIAR-FAO accords in 2006, and all Annex I PGRFA collected after the Seed Treaty's entry into force (in 2004). All those transfers are subject to the same SMTA that governs facilitated access to PGRFA provided by contracting parties. The rules on access to, and benefit-sharing from, the CGIAR collections within the Treaty's Multilateral System have important implications for the situation-structural hypothesis, an issue to which I return below.

Another question regarding the scope of the Multilateral System concerned private collections, for instance those of the plant-breeding industry. While little is known about private collections, they are likely to be substantial. Control over germplasm collections is one factor driving corporate mergers and acquisitions in the seed sector; and some private collections are possibly larger than some public ones within the scope of the Multilateral System.[62] As a Tanzanian delegate noted in 2001, "one of the weaknesses of the Treaty is that many developed countries have effectively excluded access to resources that are in private institutions and continue to privatize collections of genetic resources that were formally in public hands. This represents a serious imbalance in commitment to the Treaty, since virtually all PGRFA in developing countries are in the public domain while in developed countries more and more are being privatized and progressively being more locked into [intellectual property rights] systems."[63] Should natural and legal persons who generate commercial profits from access to the PGRFA provided by others be required to reciprocate by making their own collections available in the same manner? Legally, the answer is straightforward: as several industrialized countries pointed out, the mandatory inclusion of such collections would constitute a violation of private property rights.[64] Article 11.4 grants the Governing Body the option of denying access to natural and legal persons that do not reciprocate by placing their own collections in the Multilateral System.

As already noted in regards to the Chairman's Elements, decisions on the Multilateral System scope are a strong data point in favor of the interplay management hypothesis. The Treaty's provisions on "encouraging" private holders to place their collections in the System, as well as the option of denying access to those who do not do so, avoids inconsistencies, or even manifest conflicts, with private property rights while attempting to maximize the System's coverage. Similarly, the decision to grant facilitated access to the CGIAR collections without differentiating between parties and non-parties avoids the problem of excluding the latter from a stock of PGRFA held in -trust for the international community as a whole. Yet access by non-parties, and users within their jurisdictions, being subject to the same SMTA as parties also has bearing on explaining the decision of a majority of FAO members to forego US participation in order to maintain what is arguably the most-controversial provision of the entire Seed Treaty, namely article 12.3.(d).

The US and the article 12.3(d) controversy

While the third contact group's meeting in Tehran resulted in agreement on the linkage between benefit-sharing and intellectual property rights, its subsequent session in Neuchâtel saw Australia, Canada, New Zealand and the US backsliding on their earlier commitments and raising concerns about the provisions of the draft Treaty, specifically the later article 12.3(d), being incompatible with the TRIPS agreement.[65] This article was among the last remaining components of the Multilateral System (article 11 to 14) on which delegates did not achieve consensus at the last negotiating session before the November 2001 meeting of the FAO Conference, which was to adopt the final text.[66] When the Conference convened, the US delegation argued that article 12.3(d) would make US ratification impossible and, accordingly, requested its deletion. Delegates dismissed the motion with 97 to 10 votes.[67] The US never became a party to the Treaty; Japan initially withheld its accession yet joined in 2013.

In its final version, article 12.3(d) reads that "[r]ecipients shall not claim any intellectual property or other rights that limit the facilitated access to the plant genetic resources for food and agriculture, or their genetic parts and components, in the form received from the Multilateral System." The exact meaning of this article, and the extent to which (or whether!) it limits intellectual property claims, has never been clarified.[68] First, the general consensus is that "restrictive" intellectual property rights, for the purposes of the Seed Treaty, refers to patents and not plant variety rights; as the Breeders' Exemption under the latter arguably allows third parties the unauthorized use of protected materials for those purposes for which the Seed Treaty grants facilitated access, they are not "restrictive." That is, PGRFA from the Multilateral System, which are brought under plant variety rights, are in general available for further access.

Second, a more difficult question is what the term "in the form received" refers to. PGRFA "in the form received" (that is, without having undergone some kind of improvement through classic plant breeding or biotechnological processes) are not eligible for plant variety rights (or patents, where available) as they lack the criterion of novelty. Plant varieties and inventions that are not "new" cannot be protected in the first place, which would make the term "in the form received" redundant. A different interpretation is that the prohibition does not relate to patent claims on materials "in the form received," but rather that "no intellectual property rights can be taken out over the material, or subsequent products derived from that material, if the effect would be [to] reach back and *limit facilitated access by others to the original material* accessed."[69]

Third, the inclusion of "parts" and "components" implies that gene patents fall within the scope of the prohibition. The extent to which gene patents, rather than patents on plants or plant varieties, would limit further access is a matter of legal dispute. As discussed in Chapter 2, the patentability of isolated genes varies across jurisdictions and is also subject to evolving case law. Isolated genes "in the form received" would arguably fall under the Article 12.3(d) prohibition.[70] Before the US Supreme Court ruling in *Myriad Genetics* in 2013, this could have led to legal conflicts insofar as national law had allowed for patents on isolated genes if their functions could adequately be described, with the Seed Treaty prohibiting such claims on genes isolated from Annex I materials. The article possibly also applies to patents limiting facilitated access to the genes (the "parts" and "components") contained within Annex I materials, not just materials-as-plants or plant varieties.[71]

This data point is important insofar as developing countries willfully sacrificed US participation—the largest player in global biotechnology and a significant holder of *ex situ* PGRFA. Similar to the legal uncertainty surrounding article 12.3(d), the theoretical explanations tested here all fail to explain this design element, albeit in different ways.

Explaining the institutional outcomes

The transition from the 1983 International Undertaking to the Seed Treaty is a story of both continuity and change. Major components of the former persist in the latter in slightly modified form. The 1991 amendment incorporated the principle of state sovereignty, which the Seed Treaty adopts as well. And, while the latter does not refer to the "common heritage of mankind," it considers PGRFA as a "common concern of all countries."[72] The Seed Treaty thus reproduces the uneasy tension between the two principles rather than replacing one with the other. This tension is also apparent in its operational provisions on ABS: access to PGRFA within the Treaty's Multilateral System does not require the explicit granting of Prior Informed Consent by provider countries, the negotiation of Mutually Agreed Terms and bilateral benefit-sharing arrangements. The reason for this is the same that gave rise to the common heritage-principle under the International Undertaking: global interdependence in plant breeding.

While the Seed Treaty sports a fully fledged system for fair and equitable benefit-sharing, the Undertaking "recognized" that PGRFA are to be preserved and used "for the benefit of present and future generations";[73] resolution 4/89 decided that users may be charged for access to PGRFA, while resulting benefits are part of a "reciprocal system" and should accrue to those countries adhering to the Undertaking; resolution 5/89 endorsed the concept of Farmers' Rights to allow farmers, their communities and the world's countries "to participate fully in the benefits derived" from the utilization of PGRFA. To be sure, this is a long way from an operational regime for fair and equitable benefit-sharing as under the Multilateral System. Yet, in terms of their underlying approaches, the two regimes exhibit a high degree of continuity.

At the same time, the Seed Treaty contains various novel elements that shape its relationship to other institutions within the genetic resources regime complex. First, while Treaty as such covers *all* PGRFA, its ABS provisions only relate to the limited set of crops contained in its Annex I as well as the CGIAR collections. As discussed at length above, limiting the scope of the Multilateral System allowed provider countries to leave valuable PGRFA within the ambit of bilateral benefit-sharing under the CBD, thus closing the "loophole"[74] the International Undertaking had left within the latter. Bringing the CGIAR collections within the Multilateral System, moreover, allowed resolving their ambiguous legal status after the CBD's entry into force.

Second, the Treaty generates novel institutional overlaps with the international patent regime. Despite more than ten years of legal debate, its precise implications for the patenting of plants and their components remain unclear. Yet, unlike the Undertaking, its relationship to plant variety rights is relatively clear-cut. The Undertaking aimed to make PGRFA, including protected varieties, available "for plant breeding and scientific purposes" without restrictions.[75] This approach is incompatible with plant variety rights in line with the UPOV Convention, which only allows unrestricted access to protected varieties for the narrow set of purposes outlined in the Breeder's Exemption. The Seed Treaty resolves this problem, as facilitated access to Annex I PGRFA is exclusively granted for "research, breeding and training for food and agriculture,"[76] that is, for the same purposes set out in UPOV's Breeder's Exemption.

Considering the divergence of interests among providers, providers/users and non-users/non-providers, why did parties choose to amend the legal framework of the International Undertaking and subsequently use it as a blueprint for negotiating a new, multilateral treaty? Why did they not choose to go down the minilateral route by cooperating through clubs of like-minded actors where minimal interest differentials would have made agreement on key design features much easier? I address the first question by looking into how interplay management, regime shifting and situation structure explain the factual institutional changes. For the second question, I turn to the counterfactual: what would have been the consequences for each of the three groups of actors if it had chosen club cooperation instead?

Interplay management

The interplay management hypothesis can explain the amendments of the International Undertaking, in ways that raise an important follow-up problem. In the face of the problematic relationship between the Undertaking's open access

approach and the access restrictions imposed through plant variety rights, resolution 4/89 can be interpreted as an attempt to avoid legal inconsistencies when
UPOV members must grant proprietary rights on plant varieties yet have committed to ensuring unrestricted access to such varieties under the Undertaking.
Similarly, the imminent conclusion of the CBD can be understood as a factor
driving the recognition of state sovereignty under resolution 3/91 in order not to
lead to a normative conflict with the common heritage approach.

The problem with this explanation is that, in attempting to ensure consistent
relationships *between* those institutions, the inconsistencies within the Undertaking itself mounted. Regarding plant variety rights, the Undertaking's open
access approach is significantly broader than the Breeders' Exemption under the
different versions of the UPOV Convention. That is, the two *are* incompatible,
regardless of the content of resolution 4/89. International agreements do not cease
to be "incompatible," or inconsistent, simply because contracting parties state the
opposite. The only way to resolve this incompatibility, thus, would have been be
to narrow the scope of access under the Undertaking to the size of the window
left by UPOV's Breeders' Exemption. Resolution 3/91 similarly increased the
inconsistencies within the Undertaking itself through its haphazard introduction
of state sovereignty. Subjecting the common heritage of mankind to the sovereignty of nation states leads to inconsistency as the absence of sovereignty is
commonly understood to be a key element of the common heritage principle.[77]
While resolutions 4/89 and 3/91 can thus be understood, on one hand, as attempts
to manage the interplay with UPOV and the CBD, respectively, the concomitant
introduction of inconsistencies into the Undertaking itself makes the argument
that the objective of ensuring institutional functionality drove bargaining behavior hard to sustain.

The interplay management hypothesis faces its largest problem when attempting to explain the linkage between access and intellectual property rights. The
wording of the Seed Treaty's article 12.3.d and the equivalent provisions of the
later SMTA are so ambiguous that their meaning is still a matter of scholarly
dispute.[78] The scope of the article 12.3.d prohibition is simply unclear: what is the
precise, biological meaning of "parts and components"? Where is the threshold
between PGRFA that are "in the form received" and those that have undergone
sufficient biotechnological innovation to escape the scope of the prohibition? Do
"restrictive" intellectual property rights include all types of patents across all jurisdictions? Are states or users the subjects of the prohibition?[79] There is simply
no clear, unambiguous division of labor between the Seed Treaty and the international patent regime, jeopardizing institutional complementarity.

Interplay management fares better in regards to scope. The choice to place
the CGIAR collections within the Multilateral System promised to resolve the
former's unclear legal status in a pragmatic fashion. When the CBD entered
into force in 1993, a key issue had been who, if anyone, exerts sovereignty over
those collections? Since the 1994 in-trust agreements, the CGIAR centers were
conserving the collections for the benefit of the international community as a
whole, yet this did not resolve the question of ownership. Placing them within the

Multilateral System was one way of ensuring both facilitated access to, and fair and equitable benefit-sharing from, the wealth of genetic materials held by the CGIAR. Yet to some extent, interplay management is observationally equivalent with situation structure: the clear division of labor between the CBD and the Seed Treaty is explicable in terms of actors seeking to avoid potentially disruptive institutional overlaps, as well as in terms of situation structure, with powerful provider countries (such as Brazil and China) wishing to apply bilateral benefit-sharing to their most-valuable PGRFA.

The choice to grant facilitated access exclusively to PGRFA listed in Annex I (and the full CGIAR collections), and exclusively for agricultural ends, helped delineate the scope of the Seed Treaty in relation to the CBD, establishing a clear division of labor. Accordingly, both users and providers understand which of the two regimes applies to a particular genetic resource and what the precise consequences are. The later Nagoya Protocol returns the favor by creating an exemption for the Multilateral System under article 4.4.

The decision for "facilitated" access to PGRFA within the System, finally, takes account of the state sovereignty principle by doing away with open access while allowing for provider countries to grant their Prior Informed Consent implicitly under article 11, with Mutually Agreed Terms being standardized within the SMTA. Similarly, the choice for multilateral benefit-sharing is consistent with the CBD insofar as it takes place "fairly" and "equitably," the benchmark for those terms being Mutually Agreed Terms, in the form of the SMTA, between providers and users.

Interplay management can thus explain the institutional design of the Seed Treaty in regards to the CBD, yet cannot explain its relationship to the international patent regime. The latter can partially be explained by strategic inconsistency, which, however, falls short in other ways.

Regime shifting

Regime shifting is congruent with institutional outcomes as the international regulation of PGRFA indeed underwent significant changes. The literature has especially emphasized the Seed Treaty's intellectual property dimension.[80] Article 12.3(d) can thus be explained as a counter-regime norm intended to roll back the scope of the international patent regime. Developing countries have shown such behavior in multiple international forums where intellectual property rights are being addressed.[81] Yet the question is whether this behavior aims at undercutting international patent law at a general level or at capturing specific material gains by limiting the negative externalities that plant patents may generate for third parties, as situation structure would suggest. Another question is whether either has been a successful strategy. On one hand, it is simply not clear to what extent article 12.3(d) actually limits patent claims. On the other, it is equally unclear to what extent the article mitigates negative externalities from avoided patents. The fact that the article was so crucial for developing countries that they were willing to forego US ratification to keep the text within the Seed Treaty suggests one of two things: either the participation of the largest user country of PGRFA, which holds significant *ex situ* collections and accounts for the world's highest share of relevant patent claims was less important than the creation of a counter-regime norm or, in situation-structural terms, US participation was less critical than it appears (a point to which I return below). In any case, the relevance of article 12.3(d) for patent law is unclear, also considering that the EU interprets it as not imposing patentability requirements that go beyond existing ones. Does this shift the Seed Treaty into the domain of patent law? And: if this was the intention, why did developing countries give up the idea of including limitations on plant-variety rights, through an extensive concept of Farmers' Rights, early on in the negotiation process (see below)?

There is another difficulty here. Even assuming that article 12.3(d) limits patent claims, article 13.2.(d)(iv) explicitly *enables* patent claims on those Annex I PGRFA that escape the article 12.3(d) prohibition, with subsequent commercialization triggering mandatory benefit-sharing. While the Treaty's access component could be read as an attempt at weakening the patent regime's applicability to PGRFA, its benefit-sharing provisions do the opposite by legitimizing patent claims on improved materials and simply requiring rights holders to share parts of their commercial profits under a commercial contract without infringing on their rights to enjoy patent protection. The timing is instructive here: agreement on the text of article 13.2(d)(iv) emerged relatively quickly after discussions on patents had commenced: at the third meeting of the Contact Group in 2000, the language that would make it into the final treaty was already in place. Given that industrialized countries and major industry groups had initially proposed and subsequently sponsored the linkage between patents and benefit-sharing, the timing suggests that developing countries were perfectly satisfied in conceding patentability in return for the prospect of mandatory benefit-sharing without holding out for a better deal further down the road.

Regime shifting also faces problems in regards to Farmers' Rights, the "rearguard way to demand recognition of farmers within an increasingly dominant discursive framework that took property rights as its principal symbolic currency."[82] Looking back at the 1989 amendments of the International Undertaking, resolution 5/89 is consistent with the hypothesis, with developing countries attempting to counter the expansion of plant-variety rights and the transformation of traditional agricultural systems into industrialized ones, a process that started in the 1960s and was largely controlled by companies and scientists hailing from the Global North. The problem with this explanation is the linkage between resolutions 5/89 on Farmers' Rights and resolution 4/89 on plant-variety rights, the adoption of either having been contingent on the adoption of the other. This linkage begs the question of why developing countries would establish a norm under 5/89 with the intention of counterbalancing precisely what they simultaneously conceded under 4/89. While the former resolution is consistent with an explanation in terms of regime shifting by itself, it is inconsistent with the issue of linkage, by which the resolution came about.

The history of Farmers' Rights within the negotiations on the Seed Treaty begs further questions. Several of the key aspects of Farmers' Rights, such as international responsibility, the right of farmers to require Prior Informed Consent and to re-use farm-saved seeds, among other items, were thrown out of the window early on in the process. From 1999 onward, the issue was hardly on the agenda of the CGRFA and the Contact Group. The later article 9 thus became the first part of the later Seed Treaty, on which delegates found consensus. This choice was enormously controversial with participating actors from civil society.[83] It has been argued that the agreement on Farmers' Rights presented a moderate breakthrough for the countries of the Global South, as key components of the concept are contained not in article 9 but in other parts of the Seed Treaty.[84] Yet agreement on the later article 9 was reached before agreement on those other components could even be foreseen—possibly even before the 1999 Chairman's Elements had sketched out the rough contours of the later Seed Treaty![85]

While issues such as intellectual property rights, benefit-sharing and Annex I were bartered over until the very negotiation endgame, the shedding of Farmers' Rights of much of its normative content, within the course of a few months, suggests that it did not enjoy a high priority among developing countries when compared to the other items on the agenda. The counterfactual is thus: had Farmers' Rights mattered for rolling back the impact of intellectual property regimes on farming communities, including their right to re-use farm-saved seeds, developing countries would have kept the issue on the table in the hopes of later adopting it as part of a package deal. The final Seed Treaty thus does little, if anything, to change the applicable international approach to the ways in which farming communities may (and may not) use their seeds.

Situation structure

Interplay management can explain some aspects of the Seed Treaty's institutional design yet not others; regime shifting can explain some aspects that interplay management cannot, yet in a way that is observationally equivalent with situation structure. The situation-structural account argues that actors push for institutional changes for capturing higher gains in terms of collective goods and mitigated negative externalities as well as that inconsistencies between regimes may arise as a side effect of layering new elements onto an existing institutional core.

Understood in this manner, resolution 4/89 constituted a concession for enhancing participation in the Undertaking, specifically of those industrialized countries most active in the utilization of genetic resources. Both resolutions 4/89 and 5/89 contain embryonic commitments to benefit-sharing, thus offering developing countries the prospect of participating in the commercial gains that were increasingly being realized during the take-off of agricultural biotechnology in the 1980s and early 1990s. The incorporation of state sovereignty into the Undertaking, in turn, held the promise of better control over access to genetic resources for leveraging the sharing of benefits, which the option of charging for access to PGRFA under resolution 4/89 had already made possible.

For the situation-structural approach, Farmers' Rights only matter insofar as they allow for the realization of material gains. This explains why developing countries pushed for additional financial contributions from industrialized countries with resolution 5/89. It also explains why the former were quick to drop much of the normative content of Farmers' Rights, such as international recognition and farmers' participation in relevant international decision-making processes yet (successfully) pushed for financial support (articles 5.1(c) and 18.5) and improved access to and transfer of technology for farmers in developing countries (article 13.2(b.iii)) outside of the article 9 provisions of the Seed Treaty.

On scope, there is sufficient evidence to show that, for large provider countries of PGRFA, commercial motives were dominant: in order to maximize gains from bilateral benefit-sharing under the CBD, those countries preferred a limited, list-based approach under the Multilateral System. The haggling over Annex I is another case in point: governments made the inclusion of their PGRFA in the Multilateral System contingent on the inclusion of PGRFA by other governments, even withdrawing their proposed inclusions as a reaction to withdrawals by others,[86] and sought to limit the coverage of the System to agricultural uses to allow for bilateral benefit-sharing deals on, for instance, the pharmaceutical utilization of their genetic resources.[87] Conversely, user countries with strong innovative capacities in agricultural biotechnology repeatedly voiced their preference for broad coverage without restrictions on intellectual property claims.[88]

Situation structure can explain the linkage between benefit-sharing and patents better than regime shifting does: receiving monetary transfers from commercial users of PGRFA presupposes the availability of intellectual property rights for setting the proper economic incentives for innovation. As noted above, the compromise on allowing patent claims on improved Annex I PGRFA in exchange for mandatory benefit-sharing once commercialization takes place was made a mere three negotiating sessions after the Montreux meeting and thus way before the negotiation endgame. This trade-off suggests that the receipt of monetary transfers trumped concerns over intellectual property. It is also congruent with the idea that non-users/non-providers, as the only group for which limitations to intellectual property claims was a serious issue, had the lowest leverage in the negotiations due to their disadvantaged position in the situation structure.

What situation structure cannot adequately explain is the puzzle of article 12.3(d) and the trade-off through which it made its way into the Seed Treaty. For the explanation to hold, developing countries must have expected the gains from this article to be higher than the gains from US participation, which was contingent on the article's deletion. While an answer cannot be given with certainty, two factors suggest that US participation was, in fact, less critical than would appear based on its innovative strength in agricultural biotechnology and its large *ex situ* collections.

First, negotiators were aware that, pursuant to article 12.2, they would only be required to grant facilitated access to other contracting parties as well as natural and legal persons under their respective jurisdictions. US-based users would thus be unable to gain access to Annex I PGRFA provided by contracting parties. At the same time, US users would only access the CGIAR collections (both Annex I and non-Annex I PGRFA collected before 2004 and Annex I PGRFA collected after) under the same conditions that apply to contracting parties. Annex I PGRFA held by the CGIAR would fall under the SMTA whereas non-Annex I PGRFA would fall under an SMTA to be developed by the Treaty's Governing Body.[89] US users accessing the CGIAR collections would thus be bound by the SMTA regardless of whether or not the US became a party to the treaty. The second issue is access to *ex situ* collections managed and controlled by the US government. The US National Plant Germplasm System is one of the largest PGRFA collections in the world and grants unrestricted access to more than 570,000 accessions as of August 2016.[90] Those collections are accessible regardless of whether or not the US becomes a party to the Treaty. In fact, for those users of PGRFA, non-participation is advantageous as US collections can be accessed without incurring obligations from the Seed Treaty's SMTA.

So how much can situation structure explain how the interests of groups of actors, weighted by interdependence, translate into institutional outcomes? The hypothesis predicts that outcomes should be biased towards the interests of those actors enjoying the most-privileged overall position in the situation structure. This is the question of "who gets what"? As noted in section 5.2, "providers" of PGRFA (that is, countries with particularly large *ex situ* collections yet without significant capacities for utilization) should have the strongest orientation towards the status quo in which valuable PGRFA fall under the bilateral regime of the CBD by default. This group of actors was least dependent on a negotiated outcome, as they would have had the option of simply subjecting *all* of their PGRFA to the bilateral approach. Non-providers/non-users, that is, countries with neither substantial *ex situ* collections nor commercial plant breeding capacities, were the most reformist yet were simultaneously highly dependent on others mitigating the negative externalities from national access regulation and intellectual property rights. Users/providers, those (industrialized) countries with large *ex situ* holdings as well as strong capacities for plant breeding, were in the middle, attempting to limit the post-CBD "access chill"[91] and making a maximum amount of PGRFA easily accessible without limitations on subsequent intellectual property claims.

Drawing up scoresheets for "who got what?" in the negotiation of international treaties is notoriously difficult. Yet it is safe to say that outcomes are congruent with the interests of providers as the most-privileged actors in the situation structure, as many valuable PGRFA are excluded from the regime's scope, and for those PGRFA that are in, benefit-sharing is mandatory once recipients claim patents on improved materials. For users/providers, the Seed Treaty grants facilitated access (that is, access under conditions better than those under the CBD), to a narrow range of PGRFA provided by contracting parties, yet also to the full range of samples held by the CGIAR collected before the Treaty's entry into force, that is, both Annex I and non-Annex I. Even presupposing that article 12.3(d) limits patent claims in ways that go beyond existing patent law, Japan is the only party to the treaty that allows for patents on plant varieties in the first place. The EU, in turn, interprets the Treaty as not posing additional hurdles to intellectual property claims and thus bypasses any restrictions article 12.3(d) may or may not contain. For non-users/non-providers, finally, the Multilateral System grants access to a rather limited set of crops domestic breeders may access; simultaneously, the Treaty neither imposes clear limitations on patent claims, nor contains any restrictions on the claiming of plant variety rights. The latter point is particularly important, as those countries can choose not to allow patent protection of plant varieties yet then must instead provide for an "effective sui generis system" under TRIPS article 27.3(b). Without limitations to plant variety rights under the Seed Treaty, they failed to achieve what had been a major policy objective since the 1980s. While it is difficult to rank those outcomes, it appears that, in terms of situation structure, the first group got what it wanted, the second group mostly got what it wanted and the third group got very little. In that sense, outcomes are congruent with interests weighted by interdependence.

The club counterfactual

Given that different actors were pursuing partially incompatible policy objectives, why did they resort to multilateral cooperation, knowing full well that this would require linkages and concessions? What would have happened if each of the three groups had chosen club cooperation over institutional layering?

Providers, though seeking to limit the scope of the ABS regime for the PGRFA they provide themselves, have intrinsic interests in securing access to plant breeding materials. A "provider club" among, say, Brazil, China, Russia and India, would have allowed each of them to leave key PGRFA within the ambit of the CBD while granting each other access to less-valuable PGRFA. Yet how would they have obtained access to the PGRFA held by governments *outside* the club? Assuming that those outsiders would have concluded the Seed Treaty nevertheless, facilitated access to Annex I PGRFA under the control and management of contracting parties would have been limited to *other* contracting parties—club members would have thus been unable to obtain materials from non-members. Assuming that club cooperation would have led to the collapse of the negotiations on the Seed Treaty, the situation would have been similar: no facilitated access would have been available. By keeping tabs on their domestic collections in the multilateral setting instead, providers reconciled both goals: bilateral benefit-sharing from key PGRFA they hold as well as access to the publicly held Annex I PGRFA of all other contracting parties.

A provider/user club would have been able to grant unrestricted access to the full range of PGRFA in their respective *ex situ* collections, no strings attached. Similar to the counterfactual provider clubs, access to non-members' collections would have been prevented if the Seed Treaty had been concluded. If club cooperation had led to the failing of the negotiations on the Seed Treaty, there would still have been the possibility of accessing the CGIAR collections, albeit under legally unclear conditions and the follow-up problem of the original provider countries attempting to bring those collections into the fold of state sovereignty.

For non-users/non-providers, finally, club members could have restricted intellectual property claims and abolished national access regulations, yet this would have been inconsequential, as they only could have made available to each other marginal amounts of PGRFA. If the negotiations on the Seed Treaty had failed, access to the PGRFA of non-members would still have been hampered by access regulation, and non-members would still claim intellectual property rights within their own jurisdictions. While club members could establish rules for themselves, which abolish intellectual property claims *per se*, the follow-up problem would be compliance with the minimal standards under TRIPS. The only way in which club cooperation for this group of actors would have been remotely viable is if they had been willing to breach their international obligations under TRIPS, in all likelihood subjecting themselves to WTO dispute settlement proceedings, and still not having non-members limit intellectual property claims or cut back on access regulation.

The counterfactual suggests that club cooperation was simply unfeasible for all actors involved. The painstaking negotiations on the Seed Treaty, with all the concessions and linkages it involved, were thus ultimately superior to the alternative. While club cooperation would have achieved nothing, institutional layering achieved a little—more for those actors in privileged positions, less for those in less-privileged ones.

Notes

1 Rosendal (2000).
2 Fraleigh and Harvey (2011), 110; see also Visser and Borring (2011).
3 Shiva (1993).
4 Murphy (2007a).
5 Brand et al. (2008), 111.
6 Kloppenburg (1987), 36.
7 Ackrill and Kay (2014).
8 Greenpeace (1999).
9 FAO (2010), Chapter 2; Greenpeace (1999).
10 Frankham et al. (2004), 155.
11 FAO (2010), Chapter 3.
12 ibid., 71.
13 Chiarolla and Shand (2013), 5.
14 Parisi et al. (2016).
15 ETC Group (2013).
16 Patents on processes for the modification of plant genotypes, phenotypes and for plant reproduction; as well as for recombinant DNA and RNA technologies for genetically modifying plant cells in ways that do not occur naturally Codes A01H 1 to 4 and C12N 15/82 to 84 under the International Patent Classification (IPC); 1 January 1994 to 1 January 2002.
17 Rabitz (2015).
18 Filomeno (2014); see also Stannard (2013), 252.
19 Coupe and Lewins (2007), 19; see also Egziabher et al. (2011), 52.
20 Mooney (1983), 24–33.
21 Kloppenburg (1988), 172–174.
22 Bordwin (1985), 1064–1065.

23 FAO Conference resolution 5/89.
24 Andersen (2008), 95.
25 Resolution 4/89, preamble, para b.
26 Already in its original form, the Undertaking allows the charging of fees as parties may choose to make plant genetic resources available on Mutually Agreed Terms, see article 5.
27 Cooper (1993), 160–161.
28 Resolution 4/89, para 5.
29 Resolution 5/89, para c.
30 Cooper (1993), 161.
31 ten Kate and Diaz (1997), 286.
32 Nairobi Final Act, 1992, Resolution 3.
33 FAO Council, Resolution 7/93, 1.b.
34 CBD COP 2, decision II/15.
35 FAO Conference resolutions 4/89, 5/89 and 3/91.
36 CPGR-Ex1/94/3, 8–10.
37 C 99/PV, 93.
38 For instance, see CGRFA-Ex3/96/3, Appendixes 1 to 4.
39 Mwila 2013, 228; CGRFA-Ex3/96/Rep, App. H, Att. 1. The African Group would opt for narrow coverage in the later stages of negotiations when it became apparent that few limitations on intellectual property claims would be included in the Seed Treaty.
40 For France and the US, see CGRFA-Ex3/96/3, Appendixes 1 to 4. For Australia, see ENB (1997).
41 CGRFA-Ex3/96/3.
42 Andersen (2005).
43 ENB (1997), 5–7.
44 CGRFA-Ex3/96/Rep, appendix G.
45 CGRFA/IUND/CNT, article 12.
46 CGRFA-8/99/13, Annex, article 15.
47 Coupe and Lewins (2007), 20.
48 Coupe and Lewins (2007), 20.
49 See CGRFA/Ex-6/01/3, 2–3.
50 CGRFA/CG-3/00/TXT, 6.
51 Formerly ASSINSEL, Association Internationale des Sélectionneurs pour la Protection des Obtentions Végétales.
52 Brand et al. (2008), 120.
53 ENB (2000a); Lightbourne (2009), 501–504.
54 McManis and Seo (2009).
55 Seed Treaty, preamble.
56 CGRFA/CG-2/00/TXT, Appendix B.
57 Coupe and Lewins (2007), 29.
58 Falcon and Fowler (2002), 211.
59 Stannard (2013), 253–254.
60 ENB (2001), 4.
61 Halewood et al. (2013), 109–110.
62 Chiarolla and Shand (2013).
63 C 2001/PV, 68.
64 ENB (2001), 6.
65 Brand et al. (2008), 120–121.
66 CGRFA-Ex6/01/3, 11.
67 C 2001/PV, 70–71.
68 Moore and Tymowski (2005), 92–93; Correa (2006); Schaffrin et al. (2006); Chiarolla (2008).

69 Moore and Tymowski (2005), 92, italics i.o.
70 Correa (2006), 150–151.
71 Moore and Tymowski (2005), 93.
72 Seed Treaty, preambular text.
73 International Undertaking, preambular text.
74 C 99/PV, 93.
75 International Undertaking, article 1.
76 Seed Treaty, article 12.3(a).
77 i.e., Joyner (1986).
78 Correa (2006); Schaffrin et al. (2006); Chiarolla (2008).
79 Schaffrin et al. (2006).
80 Helfer (2009); Muzaka (2010).
81 Helfer (2009); Muzaka (2010).
82 Borowiak (2004), 522.
83 Ibid., 20–21.
84 Batta Bjørnstad (2004), 40–48.
85 Coupe and Lewins (2007), 20.
86 Falcon and Fowler (2002), 211; Coupe and Lewins (2007), 22.
87 Cooper (2002), 5.
88 Fraleigh and Harvey (2011); Visser and Borring (2011).
89 Seed Treaty article 15.1(a) and (b).
90 See http://www.ars-grin.gov/npgs/summarystats.html.
91 Fraleigh and Harvey (2011).

6 Biopiracy

In this chapter, I turn to the negotiations on a regime to counter "biopiracy," a process that led to the conclusion of the 2010 Nagoya Protocol to the CBD yet was preceded by a host of failed attempts at situating compliance measures within various patent law treaties. The Protocol itself contains a set of user measures that are mandatory for contracting parties to adopt and are meant to ensure that genetic resources utilized within their respective jurisdictions are being utilized in accordance with Prior Informed Consent and Mutually Agreed Terms. The Protocol also creates an internationally recognized certificate of compliance to be submitted to, or collected by, relevant national authorities (so-called checkpoints).

The multi-forum process leading up to the Nagoya Protocol is consistent with what the regime shifting hypothesis suggests, with developing countries attempting to create counter-regime norms in order to undercut, or delimit, the international patent regime. Yet the hypothesis cannot explain the eventual institutional outcome, which fails to include precisely such a norm. Conversely, outcomes are congruent with the collective interplay management explanation, yet the negotiation processes testify to the outcomes resulting from the *inability* of reformist actors to create problematic linkages with patent law, not their *unwillingness* to do so. The present case is the only one in this book in which the situation-structural explanation unambiguously excels: asymmetrical interdependence in favor of status quo-oriented actors and the inability of reformist actors to enter into club cooperation explain the failure of the various ambitious attempts at amending international patent law and the settlement for a relatively weak compliance regime for lack of better alternatives.

Compliance and biopiracy

The various attempts by developing countries to create an international ABS compliance regime are, on one hand, aimed at implementing the CBD's benefit-sharing objective. While the CBD allows national authorities to regulate access to genetic resources, it does not include an operational regime for ensuring that resources used in third countries are being utilized in accordance with the provider countries' access frameworks. On the other hand, the push towards a compliance regime was motivated by broader concerns regarding biopiracy, a term that is notoriously difficult to pin down. As noted in Chapter 4, one way of understanding "biopiracy" is in terms of misappropriation of genetic resources (access without Prior Informed Consent) and misuse (utilization in breach of the Mutually Agreed Terms between user and provider country).

A major difficulty lies in the question of the appropriate policy responses to misappropriation and misuse. Since the late 1990s, developing countries have put forth numerous proposals for a compliance mechanism revolving around the concept of "disclosure of origin": users of genetic resources applying for intellectual property protection of relevant inventions would need to provide evidence of the origins of the materials they use, thus allowing their compliance with the relevant ABS frameworks to be verified. Users might also be obliged to provide direct evidence that Prior Informed Consent had been granted, Mutually Agreed Terms established and that benefits are being shared. Non-compliance with a disclosure requirement could lead to patent applications not being processed; where fraudulent information is being disclosed, patents could retroactively be invalidated.

The idea of disclosure of origin faces several challenges. First, it may not always be possible to identify the provider country from which a genetic resource has originally been obtained. The utilization of genetic resources often takes place along value chains with multiple intermediaries, involving botanic gardens, gene banks, dedicated companies screening for useful biochemical compounds to be subsequently sold to pharmaceutical manufacturers and so forth. Requiring that users trace back the origin of a genetic resource through this entire chain would place an undue burden on them and might often not even be possible due to a lack of documentation.

A second challenge is the compatibility of various proposals for disclosure of origin with international patent law. A dedicated ABS compliance regime within international patent treaties would arguably violate the TRIPS article 27.1 requirement that WTO members not discriminate between different areas of technology. Requiring the fulfillment of specific criteria for obtaining patents on inventions incorporating genetic resources would constitute such a discrimination.

A third issue is whether a disclosure requirement would even be effective.[1] In several cases developing countries offered as instances of biopiracy within the TRIPS Council discussions, patent applicants *had*, in fact, disclosed the origin of genetic resources used in their inventions. Moreover, a major concern voiced by industrialized countries is that disclosure, as a pre-condition for obtaining biotechnological patents, would stifle innovation in the sector; less innovation, in turn, would reduce the size of the cake that can subsequently be shared.

Finally, there is the question of how international compliance measures need to be complemented by adequate access frameworks in provider countries. At the same time developing countries had been pushing vigorously for an international compliance regime, they placed significantly less attention on implementing their own obligation to "facilitate" access to genetic resources under article 15.2 CBD and on creating the legal, regulatory and administrative conditions that would allow users to access genetic resources from their territories without facing undue burdens and uncertainties. The neglect of this national component is a problem frequently pointed out by industrialized countries. As long as provider countries do not possess clear, national access frameworks, including transparent procedures for obtaining Prior Informed Consent and negotiating Mutually Agreed Terms, international measures for ensuring that users comply with such frameworks miss the point entirely.

Situation structure

This chapter deals with the process towards an ABS compliance regime covering the utilization of *all* genetic resources within the scope of the CBD. Accordingly, there are important overlaps with situation structures in the other cases in this book. Such a compliance regime thus potentially applies to PGRFA; to marine genetic resources originating from the territorial waters of CBD members and to influenza viruses, insofar as they constitute "genetic resources" in line with the CBD (see following chapter). I thus limit myself to the two most pertinent aspects defining the situation structure in the formation of an ABS compliance regime: biotechnological patents and the global distribution of biodiversity. Both are very rough indicators: the former includes a wealth of products that do not incorporate genetic resources, as well as processes that are not applicable to their utilization. The latter does not capture the fact that genetic resources differ in terms of their potential commercial value.

Patenting activity by geographical origin is a useful indicator of how the ownership of relevant technologies is distributed and the relative innovative strength of different countries and regions. For the time period covered in this chapter (~2000–2010), the dominance of a handful of industrialized countries in biotechnological patents has remained largely constant, with two notable exceptions. First, the overall share of the US declined strongly while that of the remaining OECD states remained at roughly the same level. Second, emerging economies such as Brazil, China and India have drastically increased their relative patenting activities. For Brazil and India, the absolute numbers are still marginal. This is different for China. Annual Chinese biotech patents have increased 15-fold since 2000, up to 1852 as of 2010.[2] This makes China the third-largest player in the world, after Japan but before Germany.

Measuring the global distribution of biodiversity is methodologically challenging.[3] One mapping exercise for vascular (higher) plants identifies the five centers of highest diversity with more than 5000 species per 10,000 km²: Costa Rica and Panama, the Colombian and Ecuadorian Andes, the Brazilian Atlantic coast, Northern Borneo and New Guinea. Fifteen other centers of high diversity (3000 to 5000 species/10,000 km²) exist in Central America, the Caribbean, Guayana, Cameroon and Guinea, the Mediterranean, the Caucasus, the Albertine Rift (covering parts of Burundi, the Democratic Republic of Congo, Rwanda, Tanzania and Uganda), in two different areas of South Africa, Madagascar, the Himalaya, South and Southeast Asia and the Northeast and Southwest of Australia.[4] All of those have high levels of vascular plants unique to the particular region. This implies that certain potentially valuable genetic resources are not available elsewhere.

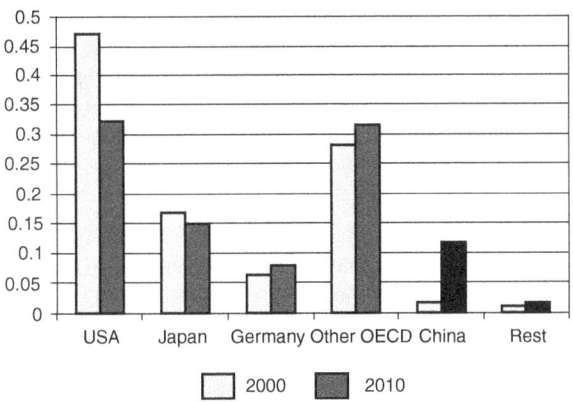

Figure 6.1 Global biotechnology patent grants by origin, 2000 and 2010
Source: WIPO Patent Statistics.

For the purposes of ABS under the CBD and related regimes, animal genetic resources presently matter significantly less than plant genetic resources. However, their global distribution is similar: the highest levels of biodiversity for terrestrial mammal species are located in South and Central America, Sub-Saharan Africa and Southeast Asia.[5] Very little is known, finally, about the spatial distribution of microbes.[6] However, large amounts of microbial genetic resources are expected to be found in the oceans, an issue to which I return in Chapter 8.

Notably, there is little overlap between the group of countries with significant biotechnological innovation and the group of countries with high or highest levels of biodiversity. The two only exceptions are Australia (ranked 13th in global biotech patent grants in 2010) and China. This clustering is what allows for the identification, common in the literature on ABS, of developing countries as (potential) provider countries and industrialized countries as (potential) user countries.

As neither the acquisition and cross-border transport of genetic resources nor the utilization of materials stored in *ex situ* collections abroad can effectively be controlled by provider countries, the cooperation of user countries is essential for ensuring the fair and equitable sharing of benefits. Yet, as in the case of PGRFA and the cases to be discussed in the following two chapters, user countries lack economic incentives for regulating the behavior of their domestic users in costly ways. At the same time, access to genetic resources depends less on cooperative arrangements than does benefit-sharing, leading to provider and user countries being asymmetrically interdependent. The situation structure thus biases user countries towards maintaining the status quo in which the utilization of genetic resources is not subject to elaborate compliance mechanisms. Conversely, provider countries are biased towards institutional change yet require the participation of user countries for any resulting compliance regime to be effective.

Better enforcement of benefit-sharing requirements is not the only type of cost associated with an ABS compliance regime. A common concern is that such a regime could impose substantial transaction costs and, if situated within patent law, lead to significant complications in processing patent applications. Finally, an issue that played a key role in the negotiations on the Nagoya Protocol was that user countries might be obliged to domestically enforce legally unclear and burdensome provider country legislation. For this reason, the EU made agreement to the Nagoya compliance regime contingent on the adoption of minimal standards for the domestic access frameworks of provider countries.

Unlike in the case of the Seed Treaty, the situation structure in the issue area of ABS compliance is relatively clear-cut: providers of genetic resources expect larger benefit-sharing flows from enhanced compliance measures whereas user countries expect larger costs for domestic industries where such benefit-sharing flows are monetary in nature.[7] This translates into opposed institutional preferences and alternatives to negotiation failure: the status quo prior to the conclusion of an ABS compliance regime in 2010 advantaged user countries as the CBD's benefit-sharing objective and article 15.1 provisions were not fully operational. Conversely, provider countries are bound to gain from the CBD's better implementation. In terms of material gains, the latter thus had more to lose in the processes leading up to the 2010 Nagoya Protocol. Its conclusion, in turn, can only be explained through the role of the issue linkages provider countries offered in order to ensure the participation of important status quo-oriented actors.

Negotiations

Negotiations on an ABS compliance instrument commenced in 2000 and spanned numerous international forums before culminating in the 2010 Nagoya Protocol. In the following, I first address the discussions under WIPO and the WTO before turning to the Protocol itself. The WIPO and WTO processes were ultimately fruitless. However, important insights can be gleamed from an analysis of the reasons for failure as well as from the linkages and concessions within those different processes.

The patent law treaty and the IGC

Discussions on a disclosure of origin requirement commenced in the World Intellectual Property Organization's (WIPO) Standing Committee on the Law of Patents in 1999. The Committee possesses a comprehensive mandate for dealing with international patent law and was about to conclude its negotiations on the Patent Law Treaty, intended to harmonize several formal aspects of member states' patent laws. At its September 1999 meeting, Colombia proposed adding language that patents on countries' "biological and genetic heritage" shall only be granted if they have been acquired legally," with applicants being obliged to provide relevant documentation.[8] While receiving support from numerous other developing countries, other Committee members rejected the Colombian proposal on the grounds that it does not relate to procedural, but rather substantive, aspects of patent law. The Committee agreed, however, to continue deliberations on genetic resources and patents under a separate negotiation stream.[9]

The following year, WIPO's General Assembly established an Intergovernmental Committee on Intellectual Property and Genetic Resources, Traditional Knowledge and Folklore (IGC) to look into three inter-related themes: access to and benefit-sharing from genetic resources, the protection of traditional knowledge, which may (or may not) be associated with such resources, and the protection of expressions of folklore as a subset of Traditional Knowledge.[10] The concept of "traditional knowledge" is notoriously difficult to pin down and goes beyond questions of ABS and genetic resources.[11] Broadly speaking, it can refer to informal knowledge that is collectively held by indigenous and local communities and forms part of their respective cultures. Traditional Knowledge associated with genetic resources may entail, for instance, the use of plants for medical or spiritual purposes in contexts where intellectual property rights, such as patents or plant variety protection, are not applicable.

Since its inception, the IGC process has made some progress on Traditional Knowledge and Traditional Cultural Expressions and virtually none on intellectual property and genetic resources.[12] One development that stands out is a 2005 EU proposal for an international disclosure of origin requirement: by amending the Patent Law Treaty, the Patent Cooperation Treaty and, possibly, the European Patent Convention, patent applications would be obliged to disclose the country of origin (or, if unknown, the direct source). Failure to do so would lead to an application's not being processed until the applicant complied with the requirement; fraudulent information would lead to consequences outside of patent law.[13] This proactive role of the EU, as well as of Norway and Switzerland (see below) is surprising at first: why is it that user countries (or a regional organization comprised of several major user countries) proffers a compliance measure from which they are bound to incur costs? In section 6.3, I return to this issue: by the 2000s, several industrialized countries had already chosen to unilaterally create "soft" disclosure requirements at the national level or were in the process of doing so. Hence, they would not have incurred costs from an international obligation to introduce such requirements. The consequence of uploading their domestic regulations to the international level would thus have been to level the playing field with those user countries *without* national disclosure requirements.[14]

At the time of writing, the IGC has failed to deliver on any of the three issues on its agenda, despite its (revised) 2009 mandate calling for "text-based negotiations" to conclude one or several international legal instruments for ensuring the effective protection of genetic resources, Traditional Knowledge and Traditional Cultural Expressions.[15] Compliance continues to feature prominently on the IGC's agenda, with the 2011 draft objectives and principles *inter alia* including the goal of ensuring that access to genetic resources and associated Traditional Knowledge be compliant with national laws, subject to Prior Informed Consent and Mutually Agreed Terms, and that patent applications disclose the origin of relevant genetic resources and Traditional Knowledge.[16] Interestingly, some evidence suggests that Brazil, one of the leading proponents of a disclosure requirement under international patent law, attempted to "remove genetic resources from the IGC's mandate in order to create additional pressure that may help to elevate genetic resources in general, and a new disclosure requirement in the patent laws in particular, to a negotiating item in the Doha Trade round" and possibly to "create pressure on the CBD Ad Hoc ABS Working Group, which is also considering a new disclosure requirement."[17]

The TRIPS review and Geographical Indications

Parallel to the founding of the IGC, debates on compliance started in the WTO. TRIPS Article 27.3.b, which allows WTO members to create exemptions from patentability, contained a provision requiring its review four years after the Agreement's entry into force; article 68 provides for the monitoring of, and consultation on, the entire agreement and the 2001 Doha Ministerial Declaration ordered the TRIPS Council "to examine, inter alia, the relationship between the TRIPS Agreement and the Convention on Biological Diversity, the protection of traditional knowledge and folklore, and other relevant new developments" that member states might raise regarding the agreement's implementation.[18] All of this quickly boiled down to two question: is there a conflict between the TRIPS Agreement and the CBD? And if there is, should TRIPS be amended?

The CBD-TRIPS relationship has been subject to heated discussions and varying interpretations.[19] On one hand, conflicts may arise from their partially incompatible objectives, as one of them aims at fair and equitable benefit-sharing, whereas the other confines the benefits arising out of the utilization of genetic resources to owners of intellectual property. The TRIPS obligation that WTO members grant patents on biotechnological inventions may also breach the principle of state sovereignty over genetic resources, as patents must be granted in accordance with TRIPS regardless of whether or not relevant biological materials have been accessed in accordance with Prior Informed Consent and on Mutually Agreed Terms. On the other hand, states that are parties to both agreements do not face obstacles in implementing their different obligations simultaneously. That is, the fact that WTO member states must grant patents on biotechnological inventions incorporating genetic resources does not prevent them from implementing arrangements for the fair and equitable sharing of benefits.[20] The EU, for instance, opines that "there is nothing in the provisions of either agreement that would prevent a state from fulfilling its obligations under both."[21] And for the US, "tailored national laws outside the patent system" would be more effective in ensuring the mutually consistent implementation of both agreements.[22]

The idea of a disclosure of origin requirement has been proposed as a solution to the alleged CBD-TRIPS conflict. By ensuring that patents can only be claimed after applicants demonstrate their compliance with the relevant provisions of the CBD (or, more specifically, the implementing legislation contracting parties to the CBD are required to put in place at the domestic level), the conflict would be resolved. The African Group presented a first, comprehensive proposal in 2003.

Under amended TRIPS article 29 (conditions on patent applicants), members would be obliged to "require an applicant for a patent to disclose the country and area of origin of any biological resources and traditional knowledge used or involved in the invention, and to provide confirmation of compliance with all access regulations in the country of origin."[23] Compliance with the CBD would be a prerequisite for obtaining patents (or other intellectual property rights) on *in situ* genetic resources as well as inventions derived from, or based on, Traditional Knowledge (see below).[24] Peru proposed an alternative scheme in 2005. Under a new article 27.3(c), members would be allowed to exclude from patentability "products or processes which directly or indirectly include genetic resources or traditional knowledge" if applicants do not comply with the applicable national and international legal access frameworks.[25]

The idea of a disclosure requirement even found limited support among some industrialized countries. In 2002, the EU had already agreed to "examine the possible introduction of a system [...] that would allow Members to keep track, at global level, of all patent applications with regard to genetic resources for which they have granted access."[26] This system would not impose additional patent-ability criteria; the consequences of non-compliance would be addressed out-side of patent law and the information to be disclosed would merely include the country of origin (where known) or source (such as a gene bank or laboratory) instead of evidence of Prior Informed Consent, Mutually Agreed Terms and/or benefit-sharing. Norway, in 2004, proposed an obligation under TRIPS appli-cable to all national, regional and international patent applications to disclose the country of origin/supplier country and, where applicable, state whether Prior Informed Consent has been obtained. Failure to comply with this requirement would lead to patent applications not being processed further, and discovery of non-compliance after a patent has been granted would not lead to its retroactive invalidation.[27] Other user countries were more reluctant. The US consistently held the line that the misappropriation of genetic resources is a problem to be addressed under national law and that international measures (including under international patent law) would not be effective in resolving the issue. Japan was similarly opposed, citing legal concerns about the TRIPS incompatibility of a disclosure requirement.[28]

In 2008, an agreement came within reach. Brazil, China, the EU, India and Switzerland, as well as the African and the African, Caribbean and Pacific Groups, submitted a proposal to the WTO's Trade Negotiations Committee look-ing to link the adoption of a disclosure requirement to an extension of the TRIPS agreement's strong protection of Geographical Indications (GI) for products other than wines and spirits and the creation of a dedicated GI-registry—both high-priority issues for the EU and a constant source of friction with the US.[29] The disclosure requirement would take the form of a formal amendment to TRIPS, and WTO members would be required not to process patent applications unless the applicant provided evidence of the country of origin or other source of genetic resources and/or Traditional Knowledge; an explicit reference to Prior Informed Consent and ABS would be subject to further negotiations.[30]

Named after the document code of the proposal, the "W/52" group quickly came to encompass over a hundred WTO members. Opposition came from, among others, Australia, Canada, Japan and the US, arguing that neither GI extension nor disclosure of origin fell within the mandate of the WTO's Doha Round.[31] High-level discussions of the linkage only took place when the WTO's Ministerial Conference resumed negotiations in 2011, after a hiatus of six years.[32] However, cracks in the W/52 group began showing in the run up to the meeting as some of its members were seen as breaking up the linkage between GI and disclosure and merely pushing forward on their respective priority issue.[33] Ultimately, the failure to conclude the package deal also tied in with the wider, structural difficulties under the WTO Doha Round and the acknowledgment, at the 2011 Geneva Ministerial, that "the negotiations are at an impasse."[34]

The PCT and the Swiss proposal

In order to reduce applicants' transaction costs, the 1970 Patent Cooperation Treaty (PCT) provides a system for obtaining a bundle of national patents via a single, international application. After a successful international patent search, the application is transmitted to those national patent offices the applicant designates. The entry into the national phase can only take place if the application fulfills the national requirements of the designated patent offices. However, under article 27.1 PCT, contracting parties are prohibited from applying national requirements that are not provided for under the PCT to international patent applications. This includes a disclosure of origin requirement. It is important to note, though, that the PCT does not limit the discretion its parties enjoy in implementing a disclosure requirement as a matter of national patent law but merely restricts the requirements they may impose on applications during the national phase entry.[35] Even where entry is permitted, national patent law may still reject applications for failing to comply with a disclosure requirement.

In 2000, a general reform process commenced for simplifying and streamlining the system. In that context, Switzerland proposed, in 2004, to "explicitly enable the Contracting Parties of the PCT to require patent applicants, upon or after entry of the international application into the national phase of the PCT procedure, to declare the source of genetic resources [...]."[36] Where applicants fail to comply with those national disclosure requirements PCT parties may have in place, those parties would be allowed to halt the processing of the application, at the national level, until the requirements are fulfilled. The Swiss proposal understands "sources" as entities that are either competent to grant access to genetic resources or to participate in benefit-sharing.[37] The term thus encompasses provider countries of genetic resources, even where those resources are stored in *ex situ* collections located abroad, as well as common pools of genetic resources such as the Multilateral System under the Seed Treaty (where the SMTA would arguably be sufficient to fulfill any relevant disclosure requirements).

While the EU was generally supportive, some developing countries pointed out, the Swiss proposal "would only permit, and not require, the introduction of laws to require disclosure of source or origin."[38] Brazil furthermore noted that disclosure should entail evidence of compliance with the applicable national ABS frameworks and that the broad definition of "source" in the Swiss proposal would be insufficient for establishing the linkage from patent applications to countries of origin.[39] In 2006, Switzerland suggested dropping the discussion in order not to unduly delay the conclusion of the PCT reform. Brazil, by then one of the most-proactive developing countries within WIPO, concurred in order to "avoid duplication with ongoing work on the same subject matter in other fora."[40]

The Swiss proposal provides an interesting data point that can be interpreted in different ways. On one hand, it is noteworthy that the initiative itself came from a user country with a strong domestic biotechnology sector that, from its position in the situation structure, should be expected to oppose stricter ABS compliance measures (or rather enable PCT member states to apply such measures to the national phase entry of PCT applications). On the other hand, the proposal would merely have prevented the national entry of PCT applications that do not disclose the source of relevant genetic resources to those countries that already possess such a requirement under their respective domestic laws.

The SPLT

Negotiations on a Substantive Patent Law Treaty (SPLT), which would further harmonize national patent laws, formally commenced in November 2000 in WIPO's Standing Committee on the Law of Patents. A project mainly driven by the EU, Japan and the US, negotiations were initially limited to questions of substantive patent law, such as definitions of prior art, novelty and inventive step/non-obviousness.[41] The linkage between patent law and genetic resources surfaced in May 2002 when the Brazilian delegation proposed that non-compliance with applicable laws on ABS be grounds for the rejection of patent applications.[42] As the Colombian proposal for including a disclosure requirement in the Patent Law Treaty was rejected for being of a substantive, rather than formal nature, the SPLT negotiations would be the proper place for addressing disclosure of origin.[43] From 2004 onward, the SPLT process began to unravel. At the Standing Committee's May 2004 session, a joint European, US-American and Japanese proposal suggested dropping disclosure of origin from the negotiation agenda in order to keep the process manageable and reduce the overall workload. Argentina, Brazil, India and Iran responded in kind and attempted to kill the proposal by raising objections on procedural grounds.[44] An informal meeting in Casablanca, in early 2005, unsuccessfully attempted to get Brazil and India on board for narrowing the scope of the SPLT negotiations. Subsequently, the 14 co-sponsors of WIPO's 2005 Development Agenda, which sought to gear the international intellectual property system more strongly towards the economic, social and environmental interests of developing countries, voiced strong opposition to the narrowing of the SPLT negotiations, preferring a broad approach with a distinct development component, including disclosure.[45] The following meeting of the Standing Committee in June 2005 has been described by the US as "illustrative of the apparent bad will and contentious nature that characterizes discussions in this committee," with a "specialized technical body of WIPO" becoming "politicized" and "detached from its core technical mission."[46] In 2006, negotiations were formally put on hold.

Towards the Nagoya Protocol

Within the CBD itself, pressure towards the better implementation of the Convention's ABS provisions had been slowly building up since the late 1990s. In 2000, a new Ad Hoc Open-ended Working Group on Access and Benefit-sharing (ABSWG) was mandated to "develop guidelines and other approaches" to serve as inputs for developing "[l]egislative, administrative or policy measures" for ABS implementation.[47] The Group's work resulted in the Bonn Guidelines, a set of non-binding, guiding principles that were adopted by COP 6 in 2002.[48] Later the same year, the Plan of Implementation of the Johannesburg World Summit on Sustainable Development called on UN members to "[n]egotiate within the framework of the Convention on Biological Diversity [...] an international regime to promote and safeguard the fair and equitable sharing of benefits arising out of the utilization of genetic resources."[49] And in 2004, CBD COP 7 mandated the ABSWG to "elaborate and negotiate an international regime on access to genetic resources and benefit-sharing" for the effective implementation of CBD Articles 15 (access) and 8.j (Traditional Knowledge).[50] This regime could include both binding and non-binding elements and compliance measures, including a disclosure of origin requirement.[51]

Negotiations picked up their pace in 2006 when a draft protocol, the "Granada Annex," emerged from the fourth meeting of the ABSWG. The Annex would serve as a basis for further discussions yet contained significant areas of disagreement. Most members of the Like-minded Group of Megadiverse Countries, the Group of Latin American and Caribbean Countries and the African Group voiced their preference for "a regime with the widest possible scope, focused on channeling benefits to countries of origin."[52] Naturally, such a regime would need to include a strong compliance component. Australia, Canada and Japan were unconvinced of the merits of a binding ABS regime.[53] Norway and Switzerland adopted a cautious "wait-and-see" approach. The EU, finally, revamped its policy positions throughout 2006 and would subsequently engage with the process in a proactive and moderately progressive manner.[54]

International Access Standards

Article 15.2 CBD obliges parties to "facilitate access to genetic resources for environmentally sound uses by other Contracting Parties and not to impose restrictions that run counter" to the Convention's objectives. However, parties have long recognized the risk of divergent and ineffective national implementation, which had been the reason for the adoption of the Bonn Guidelines in 2002. If countries adopt wildly differing and bureaucratically cumbersome implementing legislation and regulations, access is impeded and the benefit-sharing objective undermined.[55] In 2006, the EU began arguing for what would henceforth become one of its signature issues: International Acess Standards for the harmonization of national implementing measures. Those could include guidelines for and mandatory components of national access legislation; an obligation on parties to make information on national access procedures available and a commitment not to discriminate between domestic and foreign users.[56]

The EU explicitly linked those standards to measures for ensuring user compliance, including a possible disclosure requirement: if developing countries want compliance, they should be ready to undertake efforts "to ensure that national access regimes fully conform to the CBD and the Bonn Guidelines and do not discriminate against foreign users of genetic resources."[57] At ABSWG 6 in 2008, the EU further pressed for provider countries' implementation of the International Access Standards' being a prerequisite for measures to ensure user compliance. That is, "additional and more specific international commitments to support compliance with ABS requirements" cannot take place "if there is uncertainty about and a broad variety of what exactly is to be enforced in countries with users under their jurisdiction."[58]

The EU initiative thus aimed at a better balance within the future Nagoya Protocol between contracting parties' rights under article 15.1 (the authority to grant access) and their obligations under article 15.2 (facilitating access). Better access to genetic resources was the primary policy objective the EU and other industrialized country actors pursued in the negotiation process. However, the proposal stirred heated controversies with developing countries, in particular the LMMC and the African Group.[59] The Nagoya Protocol contains the International Access Standards in article 6.3, obliging contracting parties that have chosen to make access to their genetic resources contingent on Prior Informed Consent to, *inter alia*, create clear, transparent and non-discriminatory access procedures under their domestic legislation and regulations. The Protocol does not link measures to ensure user compliance to provider countries' compliance with the International Access Standards, though. Given that this linkage has been a primary policy objective for the EU, its absence can partially be explained through the non-situation structural interests the latter pursued in the Nagoya negotiations, an issue to which I will return below.[60] Despite the Protocol's user measures' not being contingent on compliance with the International Access Standards, the article 6.3 provisions constitute an important refinement and operationalization of CBD article 15.2 by creating mandatory "facilitation measures" that, moreover, are enforceable viz-a-viz contracting parties under the Protocol's article 30 compliance mechanisms and procedures.[61]

Checkpoints

The better implementation of Prior Informed Consent and Mutually Agreed Terms in user countries has been a major objective for the negotiators of the Nagoya Protocol, yet opinions strongly differed on the best ways to do so. That is, users of genetic resources would have to comply with the domestic ABS frameworks of provider countries, but how and where would their compliance be verified? The debate on "checkpoints" addressed the latter question. Checkpoints would be implementing agencies, at the national level, that receive or actively seek information documenting user compliance. As such, they would be tied to the mandatory use of internationally recognized certificates of compliance (see below).

The first discussions on checkpoints started in 2005, yet the issue only took center stage towards the final stages of the negotiation process. Once user countries, and particularly the EU, had committed to legally binding compliance components in exchange for International Access Standards, the question was not *whether* checkpoints would verify user compliance, but *which ones*. The LMMC and the African Group strongly pushed for the mandatory designation of patent offices as checkpoints or, at least, their inclusion among a list of possible checkpoints from which contracting parties would be able to choose.[62] The proposed use of patent offices as checkpoints is only explicable in terms of the patent law-based compliance mechanism, which developing countries have sought since the end of the 1990s.[63] Designating them as checkpoints would, by itself, not have made compliance with the domestic ABS frameworks of provider countries into a necessary criterion for obtaining patents on inventions incorporating genetic resources. In the same manner that all developing country negotiation groups agreed on the need for checkpoints in patent offices, all industrialized countries preferred the issue to remain off the table.[64] Ultimately, the latter had their way: the eventual Nagoya Protocol requires its contracting parties to designate checkpoints for collecting or receiving documentation on user compliance yet neither provides a mandatory or indicative list of regulatory agencies.

Monitoring

Parties to the CBD first discussed a possible certificate scheme for monitoring the transnational flow of genetic resources in 2005. An internationally recognized certificate, issued by a provider country, would allow users to demonstrate compliance abroad while minimizing transaction costs as well as creating greater transparency and legal certainty. Such a certificate is structurally similar to a disclosure of origin requirement in that it requires users to demonstrate that their utilization of genetic resources complies with the applicable national ABS frameworks. It is not necessarily a prerequisite for obtaining patents or other intellectual property rights on relevant inventions. For many developing countries, the two issues were linked:[65] provider countries would certify user compliance with domestic ABS frameworks, and users would be obliged to disclose the origin of genetic resources in relevant patent applications; should they fail to do so, patents would not be granted or, where non-compliance is detected after the fact, enforcement could take place under criminal law or lead to the revocation of granted patents.

As in WIPO and TRIPS, the search for a disclosure of origin requirement proved elusive. In fact, it continued to be one of the "hot potatoes" until the negotiation endgame.[66] The Protocol's certificate system is not linked to patent law but instead to the article 6 International Access Standards. Parties to the Protocol are required to issue permits, in line with their article 6 implementing legislation, as evidence that Prior Informed Consent has been granted and that Mutually Agreed Terms have been established, as well as to register those permits with an international Clearing House, at which point they become internationally recognized certificates of origin. Those, in turn, may be used to demonstrate compliance at checkpoints in user countries.

Scope

As the Nagoya Protocol provides a compliance regime for benefit-sharing from the utilization of genetic resources, a crucial issue was which *types* of genetic resources it applies to, when those must have been collected to fall into the Protocol's scope, and whether resources from areas beyond national jurisdiction would be included. I discussed the Protocol's relationship with PGRFA in the last chapter, and its relevance for pandemic influenza viruses and marine genetic resources will be covered in the two subsequent ones. For this reason, I will only refer to the debate over temporal and economic scope (derivatives) here.

Clearly, the Protocol would apply to genetic resources accessed after its entry into force. Yet what about access that took place before, either after the CBD's entry into force in 1993 or before that? What complicates finding an answer to this question is that, curiously, neither the Convention nor its Protocol defines the meaning of the term "access." It may be understood either as identical to utilization or as the physical acquisition of biological materials. This distinction is not trivial. For instance, were the Protocol to apply exclusively to genetic resources accessed after its entry into force, interpreting access as acquisition would imply that all resources placed in *ex situ* collections prior to this date would fall outside of its scope. Conversely, understanding access as utilization means that the Protocol applies to all *ex situ* genetic resources utilized in biotechnological processes after the entry into force, regardless of their specific date of acquisition.[67]

Another tough nut to crack was derivatives, which the Protocol defines as "naturally occurring biochemical compounds resulting from the genetic expression or metabolism of biological or genetic resources," even if they "do not contain functional units of heredity."[68] The term "derivative" had been under discussion in the CBD since at least 2002 and refers to a "continuum of very general to very specific concepts."[69] Considering that different types of biotechnological innovation processes transform genetic resources in different ways, the key question the term "derivative" addresses is after which degree of transformation the benefit-sharing obligation ceases.[70] For instance, if the information encoded within a genetic resource is used for the artificial synthesis of a modified micro-organism, would benefits still need to be shared, despite the fact that the genetic resource from which the novel micro-organism has been "derived" has not been physically accessed and has not undergone any biotechnological modification? This question is of particular importance for novel techniques of synthetic biology.[71]

The Nagoya Protocol avoids giving clear answers to the questions of both temporal scope and derivatives. In fact, the Protocol does not mention its temporal scope at all! One interpretation is thus that, by default and in line with the Vienna Convention on the Law of Treaties, the Protocol only applies to genetic resources accessed after its entry into force.[72] Yet, the Protocol includes within its scope genetic resources that fall under the CBD's article 15, access to which is thus subject to Prior Informed Consent and Mutually Agreed Terms. And, while derivatives are defined in article 2(e) and included in the definition of biotechnology in article 2(d), they are not referred to in article 3 on scope. Quite different views exist, accordingly, on the extent to which the Protocol covers derivatives.[73] Temporal and economic scope is where the Protocol is arguably at its most ambiguous. This casts doubt on the interplay-management explanation as it leaves open the extent of innovative activities in the biotechnology sector, which user countries are obliged to subject to compliance measures.

The ABS-biodiversity linkage

The Nagoya Protocol was not adopted in isolation. In fact, genetic resources and ABS are but one component of policy-making activities under the umbrella of the CBD, which was originally intended exclusively as an instrument for the protection of biodiversity.[74] The two areas are largely addressed in isolation and along separate negotiation tracks. There is, indeed, no compelling functional reason why fair and equitable benefit-sharing should be related to the conservation and sustainable use of biodiversity. Yet, as with the CBD, where developing countries made their consent dependent on the inclusion of the benefit-sharing objective and state sovereignty,[75] the adoption of the Nagoya Protocol was linked to the CBD's Strategic Plan for Biodiversity 2011–2020, including the 20 Aichi Biodiversity Targets and the Strategic Plan for Resource Mobilization, intended to provide financial and technical support for biodiversity protection in developing countries.[76] At the Nagoya summit, the G77/China made clear that all three would need to be adopted as part of a larger package deal.[77] While industrialized countries were not *demandeurs* on the Protocol, some of them placed high priority on the biodiversity component of the deal. This, again, highlights the role of issue linkage for ensuring the participation of non-excludable actors oriented towards the institutional status quo.

Explaining institutional outcomes

The Nagoya Protocol grew out of attempts lasting more than a decade at developing an ABS compliance regime in order to implement the CBD's benefit-sharing objective and article 15.1 provisions. This outcome resulted after proposals for amending various international patent law treaties had repeatedly failed. Not only does the Protocol provide a compliance regime for all genetic resources within the scope of article 15 of the CBD, but, importantly, it also contains obligations relevant for the domestic legislation of provider countries by requiring them to bring their respective access frameworks in line with the minimum standards set out in article 6. The Protocol's significance for the wider genetic resources regime complex goes beyond the question of compliance, though; the following two chapters will return to its relationship to pandemic influenza viruses and the possible creation of a global, multilateral benefit-sharing mechanism for marine genetic resources. Purely in terms of its compliance provisions, though, the most notable institutional outcome is the absence of patent-related language, despite a disclosure of origin requirement having been the holy grail of developing countries across multiple international forums for about a decade. Arguably, the Protocol falls short of the deal that had been on the table in the WTO in the context of the proposed GI-disclosure linkage. While thus avoiding problematic interactions with the international patent regime, it begs the questions of why provider countries settled for a deal that fell short of the demands they had been voicing for years; it also begs the question of whether to explain this institutional outcome in terms of interplay management or situation structure.

Interplay management

The Nagoya Protocol is considered a prime example of interplay management.[78] Indeed, its articles 4 and 8 provisions clearly delineate the CBD/Nagoya regime from both the Seed Treaty and the Pandemic Influenza Preparedness Framework; article 10 obliges parties to consider creating a multilateral benefit-sharing regime for genetic resources utilized in a transboundary context or for which Prior Informed Consent cannot be granted or obtained, with implications for marine genetic resources (see Chapter 8). I return to those issues in the following two chapters. Here, suffice it to say that the relationship between the Protocol and the Pandemic Influenza Preparedness Framework is significantly more problematic than it might appear.

The Protocol avoids messy overlaps with the international patent regime as it does not contain patent-related language and gives contracting parties broad leeway in how they choose to implement their obligations domestically.[79] This outcome is congruent with the interplay management hypothesis. Once we look at the negotiation process, though, it becomes clear that institutional outcomes did *not* result from "conscious efforts [...] to improve institutional interaction and its effects."[80] Instead, major actor groupings made conscious, but ultimately futile, efforts at including institutional design elements that would have led to highly problematic consequences for international patent law. The sheer amount of political capital spent on the various processes in the WTO and WIPO suggests that, for a substantial amount of relevant actors, the improvement of institutional interactions was a secondary objective, at best. Another data point against interplay management is the willingness of several industrialized countries to go further than they did under the Protocol, given that they would receive something in return: the proposed GI-disclosure linkage in the WTO suggests that opposition to the Nagoya Protocol's being linked to patent law was not a question of concerns over institutional inconsistency, but rather one of expediency.

The WTO package deal would in all likelihood have led to a non-processing of patent applications that do not disclose the origin, or source, of relevant genetic resources. The willingness of major user countries to consider such a deal in the first place makes their unwillingness to consider patent offices as mandatory checkpoints appear to be a question of pragmatic gain-maximization instead of a diffuse commitment to institutional complementarity.

The same goes for the Protocol's article 6 provisions on International Access Standards, which are congruent with attempts to improve the implementation of pre-existing commitments, namely, CBD article 15.2. Yet that access were highly controversial and only grudgingly accepted after the EU offered the prospect of legally binding compliance measures in return. What explains the adoption of the access standards is thus the linkage: provider countries would bring their domestic legislation in line in exchange for user countries' taking measures to ensure that genetic resources are being utilized in their respective jurisdictions in line with the applicable access legislation. While the outcome can be explained in terms of interplay management, the processes leading to those outcomes cannot.

Another important data point against interplay management is the Protocol's temporal and economic scope. The extent to which the Protocol creates obligations for user countries to ensure that genetic resources collected and/or utilized after the CBD's, but before the Protocol's, entry into force is unclear. Similarly, the Protocol provides a definition of derivatives yet does not explicitly include them within its scope. The extent of user country obligations to implement compliance measures is thus unclear.

Regime shifting

A regime shifting explanation renders exactly the opposite picture. The vehemence with which a broad coalition of developing countries has been pursuing a disclosure of origin requirement in patent applications for many years and across multiple international fora suggests that shifting ABS governance into the patent regime was a major factor motivating behavior, yet the failure to actually *create* the link to patent law shows that, plain and simple, the approach cannot explain those institutional changes that *did* take place. Whereas interplay management is congruent with institutional outcomes yet fails in explaining the processes leading up to them, regime shifting adequately explains bargaining behavior yet is incongruent with the outcomes.

The adoption of the Protocol poses an additional challenge for regime shifting: if the ultimate purpose of provider countries' engagement in the CBD, WIPO and WTO negotiations had been the link to the international patent regime, why did they settle for less? Provider countries were not *demandeurs* on the Protocol's access component or on the other parts of the broader package agreed upon at the Nagoya summit. Instead, the Protocol's compliance provisions were the only tangible pay-off they received from the entire process. The Protocol's added value for those countries lay not in its impact on the patent regime but rather in its promise of better implementing the transfer of benefits from user to provider countries.

Situation structure

So far, interplay management can explain outcomes but not processes, whereas regime shifting can explain processes but not outcomes. From the situation-structural perspective, the participation of countries generating large amounts of benefits, the fair and equitable sharing of which is to be secured through an international compliance regime, is critical. This explains the choice of fora: all initiatives for creating a compliance regime took place within multilateral settings in which major users participated—the choice of TRIPS or the different WIPO treaties here offered additional advantages as, unlike in the CBD, the US participates in those. This already shows the limits to club cooperation: a club would have failed to secure the adoption of compliance measures in those jurisdictions where utilization takes place. Only by securing multilateral rules that oblige user countries to change their domestic legislation and regulatory frameworks would benefit-sharing be implemented. The final outcome, in the form of the Nagoya Protocol, did not lead to patent law-based compliance measures, yet it is "better than nothing" in terms of securing benefit-sharing flows.

The formation of an ABS compliance regime highlights both the role of issue linkages for securing the participation of actors with strongly differing interests, as well as the trade-off between participation and depth. User countries are status quo-oriented in regards to compliance: they lack incentives for imposing potentially expensive and intrusive control and sanction mechanisms on their domestic industries.[81] Linkages are thus essential for ensuring their participation. As noted above, the proposed GI-disclosure linkage would possibly have gone beyond the Nagoya Protocol in terms of regulatory impact. Here, the EU, Switzerland and others were willing to introduce a comparatively stronger comprehensive compliance mechanism in order to obtain agreement on a priority issue of their own. In the CBD, the EU's initial offer to consider a legally binding compliance regime was contingent on the linkage with the International Access Standards. As developing countries had fewer opportunities for creating favorable package deals under the CBD than in the WTO, it is no surprise that the final Protocol lacks regulatory stringency.

At the same time, there were important differences in bargaining behavior between the members of the JUSCANZ group (Japan, US, Canada, New Zealand) on one hand, and the EU, Norway and Switzerland on the other, which show how interests are not fully endogenous to situation structure. The former group generally pursued a hard line approach towards an international ABS compliance regime, disputing the very necessity of an international, legally binding agreement.[82] Conversely, the latter have frequently shown their principled willingness to make limited concessions towards the demands of provider countries. Why is this the case?

One explanation relates to diplomatic factors. Norway has traditionally held a strong normative commitment to cooperation with the countries of the Global South. Switzerland equally has a long-standing commitment towards multilateralism, neutrality and mediation.[83] The same goes for the EU, which prides itself on its vanguard role in global environmental governance and sought to avert a failure of the Nagoya Summit after it had been sidelined at the disastrous 2009 Copenhagen climate conference.[84]

Yet there is another explanation for why those actors might have been moderately accommodating of provider countries' demands. When the Protocol was being negotiated, Norway, Switzerland and several EU members states already had already put in place, or were in the process of doing so, different forms of disclosure requirements in the absence of an international obligation to do so.[85] The 1999 EU biotechnology directive encourages disclosure in patent applications as a voluntary measure, and parts of the European Parliament had pushed for mandatory disclosure when the directive was being drafted.[86] Those disclosure regimes are considerably weaker than developing countries' proposals in the WTO, WIPO and CBD. Their origins are domestic in nature, yet they imply that an international obligation to implement certain "softer" forms of a disclosure requirement would not require significant modifications to those legal and regulatory frameworks that are already in place nationally. For those countries that already have domestic disclosure requirements, the creation of an international obligation thus holds the prospect of multilateralizing their own legislation in order to create a level playing field with those major user countries that do not yet require disclosure.[87]

This latter point aligns well with the situation-structural approach: an international obligation to implement domestic disclosure requirements does not incur additional costs for those countries that already have them in place. Understood from this perspective, situation structure renders a comprehensive explanation of the long path towards an ABS compliance regime: provider countries require the cooperation of user countries for securing benefit-sharing. User countries that faced low adaptation costs from an international compliance regime used the opportunity to seek favorable issue linkages for themselves. The Nagoya Protocol, finally, constituted a second-best option for securing benefit-sharing flows to provider countries, whereas the inclusion of International Access Standards as well as the broader Nagoya package facilitated the consent of user countries through issue linkage.

The club counterfactual

The Nagoya Protocol obliges user countries to ensure that their domestic industries share potentially extensive benefits, whereas provider countries get compliance measures that stay far behind what they had been trying to accomplish under international patent law for many years. What would have been the prospects for user and provider clubs, respectively? Compared to the case of the Seed Treaty, the answer is relatively simple.

A user club would have allowed its members to access each other's genetic resources without being obliged to monitor utilization and enforce benefit-sharing in cases of misuse or misappropriation. Regardless of whether or not the negotiations on the Nagoya Protocol had failed, the problem of "access chill"[88] would have been unresolved: providers would have had no incentive to allow club members access to genetic resources within their respective jurisdiction, given that they could not have expected the sharing of benefits. While *in situ* genetic resources could still have been accessed without the provider country's consent or knowledge, this is exactly the same as it is in the factual world.[89] Without giving providers incentives to grant access to their genetic resources under Prior Informed Consent and Mutually Agreed Terms, a user club would have allowed unrestricted access to club members' collections, but not much else.

A provider club would have suffered from the lack of compliance measures in user countries. In order to obtain benefit-sharing flows, utilization must be regulated in the jurisdiction where it takes place. Just as in a user club, a provider club could not have prevented illegal access. Yet, at the same time, providers granting Prior Informed Consent and establishing Mutually Agreed Terms with users would have foregone the Nagoya Protocol's user measures and instead rely on users' unspecific obligations under the CBD only. Given that the very reason for negotiating the Nagoya Protocol was the lack of incentives to ensure compliance on the user side, a club would have failed to improve on the status quo. Neither for providers nor users was club cooperation a feasible option at any time.

Notes

1 Hoare and Tarasofsky (2007); Carr (2008).
2 WIPO patent statistics, see http://ipstats.wipo.int/ipstatv2/index.htm?tab=patent.
3 Helm and Hepburn (2014), 9–13.
4 Barthlott et al. (2005).
5 Olson et al. (2001).
6 Green and Bohannan (2006).
7 Even many 'non-monetary' forms of benefit-sharing, such as joint ownership of intellectual property, are costly to users.
8 SCP/3/10.
9 SCP/3/11, paras 204 and 207.
10 WO/GA/26/6, para 20.
11 Dutfield (2004), 91–96.
12 Vivas-Eugui (2012), 23.
13 WIPO/GRTKF//11.
14 Falkner (2007).
15 WO/GA/38/20, para 217.
16 WIPO/GRTKF/IC/18/9.
17 05Geneva1647, para 10.
18 WT/MIN(01)/DEC/1, para 19.
19 See Rosendal (2006) for an extended discussion.
20 IP/C/W/368, 2–6.
21 IP/C/W/254, 2.
22 05Geneva2798.
23 IP/C/W/404, 6.
24 ibid., Annex.
25 IP/C/W/447, 13.
26 IP/C/W/383, para 51.
27 IP/C/W/473.

28 Carr (2008), 144–148.
29 Josling (2006).
30 TN/C/W/52.
31 ICTSD (2008).
32 The 2009 Ministerial Conference was not a negotiating session but merely intended to fulfill the legal requirement for meetings in two-year intervals.
33 IPW (2011).
34 WT/MIN(11)/11.
35 de Carvalho (2005).
36 PCT/R/WG/4/13, 1.
37 PCT/R/WG/6/11, 1.
38 PCT/R/WG/5/13, para 136.
39 PCT/R/WG/6/12, para 87.
40 PCT/R/WG/9/8/, paras 13 and 15.
41 SCP/4/2, para 9.
42 SCP/7/8, para 200.
43 SCP/9/8, para 69.
44 SCP/10/11, para 10.
45 IPW (2005).
46 05Geneva1648.
47 Decision V/26, para 11.
48 See Tully (2003).
49 Plan of Implementation of the World Summit on Sustainable Development, para 44, o.
50 Decision VII/19, section D, para 1.
51 Decision VII/19, Annex.
52 ENB (2006a), 9.
53 Ibid.
54 Oberthür and Rabitz (2014).
55 Wallbott et al. (2014), 38–39.
56 UNEP/CBD/WG-ABS/6/INF/3, 30.
57 UNEP/CBD/WG-ABS/5/INF/2, 72.
58 UNEP/CBD/WG-ABS/6/INF/3, 28.
59 ENB (2008a).
60 Oberthür and Rabitz (2014).
61 Greiber et al. (2012), 101.
62 i.e. ENB (2010).
63 Buck and Hamilton (2011), 54, fn. 34.
64 Wallbott et al. (2014).
65 i.e. UNEP/CBD/WG-ABS/6/INF/3, 8–10 and 34–35; UNEP/CBD/ABS/7/INF/1/Add. 1, 4–5.
66 Wallbott et al. (2014), 48–49.
67 Wallbott et al. (2014), 37.
68 Nagoya Protocol article 2(e).
69 UNEP/CBD/WG-ABS/7/2, paras 19 to 40.
70 Tvedt and Schei (2014), 24.
71 Bagley and Rai (2013).
72 Buck and Hamilton (2011), 57.
73 Buck and Hamilton (2011), 56–57; Greiber et al. (2012), 66–68; Bagley and Rai (2013); Tvedt and Schei (2014).
74 Rosendal (2000).
75 Rosendal (2000), 95–97.
76 Oberthür and Pozarowska (2013), 113.
77 Aubertin and Filoche (2011), 53.

78 Oberthür and Pozarowska (2013).
79 Rabitz (2015).
80 Oberthür and Pozarowska (2013), 103.
81 Rabitz (2015).
82 Wallbott et al. (2014), 41.
83 Hufty et al. (2014).
84 Oberthür and Rabitz (2014).
85 Hufty et al. (2014); Rosendal and Andresen (2014); Coolsaet (2015), 373–374.
86 Directive 98/44/EC, recital 27; Leskien (1998).
87 This pattern can be found in related issue areas as well, see Falkner (2007).
88 Fraleigh and Harvey (2011).
89 Rabitz (2015).

7 Viral genetic resources

Pandemic influenza is one of the gravest threats to global public health. The 1918 Spanish Flu is estimated to have caused between 50 and 100 million casualties worldwide. The Asian and Hong Kong Flus in the late 1950s and 1960s each resulted in a death toll of at least one million. More recently, the 2006/2007 Avian Flu and the 2009/2010 Swine Flu highlighted the risks to public health arising from the growth in international travel and the ease with which pathogens may spread across the globe.[1] Thus, "[o]f all communicable diseases, pandemic influenza probably remains the most feared by politicians, policymakers, and health practitioners alike."[2]

Influenza response requires the quick production and distribution of sufficient amounts of vaccines. The production of those vaccines requires access to candidate viruses collected from the field. Depending on the legal point of view, those viruses constitute genetic resources and thus imply an obligation to share benefits—such as vaccines. Negotiations on an ABS regime for pandemic influenza viruses commenced shortly after Indonesia stopped its forward transfers of field viruses to the WHO during the 2006/2007 H5N1 epidemic. Asserting "viral sovereignty,"[3] Indonesia and other developing countries claimed the applicability of the CBD, including the right of the provider country to determine the conditions of access. This decision led to an outcry among industrialized countries and public health experts alike. While many countries, including the US, had come to recognize the need to improve access to vaccines for developing countries,[4] they preferred doing so through voluntary donations and stockpiling and rightly considered access regulation to influenza viruses as a danger to global public health due to the delays it would cause to vaccine production. As both the H5N1 and the more recent H1N1 experience have shown, governments prioritize the acquisition of vaccines for their domestic populations as a matter of national security. The capacities for vaccine production are concentrated in a handful of countries, and wealthy countries make use of Advance Purchase Agreements to gain priority access to influenza vaccines in case of a pandemic threat.

The 2011 Pandemic Influenza Preparedness Framework builds upon the Global Influenza Surveillance Network, which had remained unchallenged for decades. The Framework utilizes a two-tiered structure of SMTAs that regulate the flow of viral specimens into and out of the network of WHO-designated laboratories.

SMTA 1 encourages laboratories in the network not to seek out patents over shared materials, whereas SMTA 2 obliges commercial manufacturers to choose among several different options for benefit-sharing. While reaffirming state sovereignty over genetic resources in general terms, the Framework leaves open the question of whether virus specimens, in fact, constitute such resources. In that sense, the applicability of the CBD and particularly of the Nagoya Protocol is a matter of debate.

My three hypotheses explain those institutional changes in different ways. From the perspective of interplay management, the well-known difficulties developing countries face in procuring adequate amounts of vaccines raise the question of how to improve access to vaccination without jeopardizing the functionality of the patent regime and the WHO's International Health Regulations. For regime shifting, asserting sovereignty over viral genetic resources offered a powerful tool against "an unfair status quo whereby lower-income countries were expected to share viruses and then rely on ad hoc charitable donations."[5] As I discuss below, though, whether or not state sovereignty actually applies to viral materials is a tricky question—as attempts to formally include pathogens in the scope of the Nagoya Protocol had failed and the text of the Pandemic Influenza Preparedness Framework is quite ambiguous, the regime shift is arguably less than complete. More than in any other case discussed in this book, the explanation overlaps with the situation-structural one. Here, institutional change was driven by epidemiologically vulnerable countries leveraging access to virus specimens collected within their territories in order to obtain larger (and predictable) amounts of vaccines during a pandemic outbreak.

Pandemic influenza and the Indonesian controversy

An influenza pandemic may arise when type A influenza viruses, which are able to infect both humans and certain animals, mutate, become infectious for humans and are able to be efficiently transmitted between humans.[6] Pandemic influenza possesses a very specific threat profile. On the most basic level, it is caused by viral infections. Unlike bacteria or fungi, viruses are highly susceptible to mutation as their genomes replicate in an error-prone manner. This high mutation rate increases the chance of a novel strain being able to bypass the human immune system. The high adaptivity of influenza viruses implies that "exposure to one viral strain does not ensure an effective immune response to others if there have been significant changes to viral surface proteins, and the emergence of influenza of a different subtype will sometimes cause severe disease because the immune system will not be able to respond effectively."[7] No broad-spectrum antiviral drugs, comparable to antibiotics or antifungal drugs, are presently on the market. Developing targeted antivirals is complicated by the rapid mutations in the viral DNA/RNA (antigenic drift) and the possibility of multiple viruses combining to form a novel strain (antigenic shift). Moreover, as viruses replicate within cells, developers of antiviral drugs need to take precautions not to damage the human host.

In addition to rapid mutations, the high contagiousness of pandemic influenza viruses poses additional challenges. Influenza pandemics may arise when viral strains from animal reservoirs, such as birds and pigs, mutate and become transmissible between humans via aerosols or droplets. This particular transmission pathway makes broad viral spread within human populations more likely than in the case of, say, viruses transmitted through direct contact with body fluids (i.e., Ebola) or disease vectors such as mosquitoes (i.e., Dengue). The high contagiousness of pandemic influenza, in turn, implies that the time window for developing and mass-producing adequate drugs is quite narrow. Combined with the difficulties of developing targeted antiviral drugs, this makes vaccination the primary pharmaceutical response strategy to pandemic influenza outbreaks.

To facilitate the influenza response, the WHO's 1952 Global Influenza Surveillance Network (GISN) of research laboratories aimed at "the collection of epidemiological data and viral isolates for early detection of epidemics and rapid identification of virus variants."[8] National Influenza Centers would forward viral specimen to WHO collaborating centers, which isolate and classify the strains and share the resulting information with vaccine manufacturers.[9] Starting in 2005, concerns arose regarding patent claims from both commercial vaccine manufacturers and laboratories within GISN on inventions incorporating viruses shared through the system. As one particularly outspoken critic put it, "[t]he fact that some of these patent claims were made by WHO GISN labs themselves [...] showed WHO's lack of interest in preventing predation of GISN's public health goods by private interest."[10] Due to purchasing power differentials, developing countries were unable to procure resultant vaccines in case of an influenza outbreak.[11] Even if they were, the Advance Purchase Agreements between manufacturers and industrialized countries implied that the limited global supply of vaccines during a pandemic would not necessarily be distributed on the basis of epidemiological criteria.

At the apex of the avian influenza epidemic in early 2007, which had its epicenter in Southeast Asia, Indonesia suspended cooperation with GISN and prohibited the export of the H5N1 specimen without the express permission of the ministry of health. The decision extended to the Indonesian branch of the US Naval Medical Research Unit Two, which monitors infectious diseases in Asia, with future shipments to take place only under a Material Transfer Agreement.[12] The decision was shaped by "the commercial interests of Indonesia's state-owned pharmaceutical companies, frustration over Indonesia's failure so far to benefit from research into the Indonesian H5N1 strain, and concerns about Indonesia's ability to pay full retail price for the millions of AI [Avian Influenza] vaccine doses it hopes to stockpile."[13] Indonesia also claimed that H5N1 virus samples fell within the scope of the CBD and would accordingly be subject to the principle of state sovereignty, with access requiring Prior Informed Consent and Mutually Agreed Terms. The controversy was stirred by Indonesia's reference to official WHO guidelines, which back then stated that "[t]here will be no further distribution of viruses/specimen outside the network of WHO Reference Laboratories without permission from the originating country/laboratory" and were subsequently removed.[14]

The conformity of Indonesia's decision with international law was far from certain. The WHO's revised International Health Regulations (IHRs) created reporting obligations on WHO member states for public health emergencies of international concern. Yet neither had the IHRs already entered into force in early 2007 nor is it clear that those reporting obligations extend to the transfer of virus specimen and associated genetic information.[15] Equally unclear was whether the CBD's principle of state sovereignty could be invoked, with viruses replicating in human cells yet human genetic resources being expressly excluded from the CBD's scope.[16]

While several other developing countries, including Brazil and Thailand, supported Indonesia's decision,[17] industrialized countries voiced serious concerns regarding the repercussions for global public health. In the view of the US, "[w]ithout full virus sequencing, Indonesia and the world will be blind to possible virus changes that could make a pandemic more likely."[18] Ultimately, no other developing countries followed Indonesia's example. Still, as the European health commissioner Kyprianoú acknowledged, concerns persisted over a "domino effect" that could jeopardize global pandemic response in its entirety.[19] Yet from the outset, industrialized countries were hostile to the idea of a novel agreement on virus-sharing; as one EU diplomat was quoted, "we need their virus, they need our vaccine, nobody needs this framework."[20]

Situation structure

Due to its transmission pathway, highly pathogenic influenza strains are not localized to a specific ecological context but rather pose global risks: "of all the recurrent human respiratory virus infections, influenza causes the worst symptoms in substantial numbers of people, of all ages, everywhere."[21] Beyond its repercussions for human health, pandemic influenza also stymies the production of and trade in poultry, birds and other animals and causes disruptions to international travel.[22] While the disease may, in principle, originate anywhere in the world, its zoonotic origins imply that regions in which humans and animals are in close proximity with each other under inadequate sanitary conditions are more likely to be the source of pandemics than others. This is especially the case in Southeast Asia, which was the epidemiological center for many of the 20th- and 21st-century influenza epidemics and pandemics. The relatively large ability of this region to supply public "bads" to the rest of the world, in turn, implies that its participation in global influenza surveillance and response is of the utmost importance.

Influenza response crucially depends on the production of and access to vaccines. The capacities for sequencing and vaccine production are highly asymmetric. Six manufacturers (Merck & Co, GlaxoSmithKline, Novartis, Pfizer, Sanofi Pasteur and Sanofi Pasteur MSD) presently account for 80% of the global vaccine market.[23] Generally, vaccine development requires large capital investments. Compared to chemically engineered drugs, "vaccine manufacturing is less standardized and less predictable" as it "often involves the complex transformation of live biologic organisms into pure, active, safe, and stable immunization components."[24] Whereas several emerging economies have been rapidly building up their pharmaceutical industries in recent years, their innovative capacity is comparably low, with business models largely hinging on the production of generic medicines. For vaccines, this problem is even more severe as the requisite technological infrastructure for biologics is more demanding than for chemically engineered drugs. As pandemic influenza viruses are highly infectious, laboratories must, moreover, possess adequate biosafety levels. Finally, the production of custom-tailored influenza vaccines over a short time complicates their reverse-engineering in order to be able to produce domestically. Access to vaccine formulations, which may be protected with patents, is thus essential for generics manufacturers.

As in the other chapters in this book, a small set of countries accounts for the largest share of patents. In the aftermath of the H5N1 and H1N1 episodes, patent applications for inventions incorporating those viruses, their components or derivatives, skyrocketed. 72% of those patents are held by inventors located in the US; rank 2 is occupied by Switzerland with a mere 8%. The only developing countries from which residents obtained patents are China and India, jointly accounting for 6% of the global total (see figure 7.1). In addition to patents on genetic materials and vaccines, technologies for the production and delivery of vaccines are usually proprietary as well. This includes adjuvants, which improve the immunological effect of an antigen; novel technologies for quickly manufacturing vaccines in large amounts, for instance, through cell cultures or the delivery of live attenuated vaccines through nasal spray.

The supply of influenza vaccines by originator companies in industrialized countries is critical for other countries to safeguard their public health. The problem of access to essential medicines is well known and goes beyond the relatively narrow case of pandemic influenza.[25] Asymmetrical abilities to procure vaccines result, first, from purchasing power differentials, which are themselves linked to patent protection. As with certain drugs, influenza vaccines may simply be unaffordable for large parts of the developing world. While this is a general

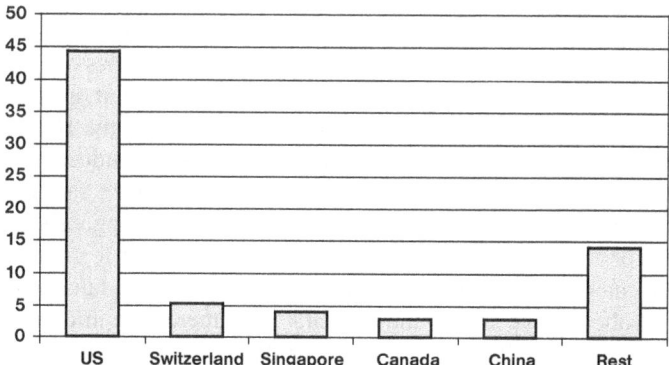

Figure 7.1 Patent grants on pandemic influenza preparedness materials
Source: WIPO, 2011.

issue in global public health, the problem of access is exacerbated by the security implications pandemic influenza poses.[26] The tightening linkages between health and security policy imply that "[s]ecurity is pursued not just through the traditional institutions of the military and police, but also via the increased development, stockpiling and deployment of new *medical* countermeasures."[27] With global vaccine production capacities being insufficient to cover the entire world population during the time frame in which a pandemic plays out, states are increasingly seeking to ensure privileged access for their own citizens. As the H5N1 crisis demonstrated, "[t]he outcome of this trend served to exacerbate the tensions between wealthier countries that could afford to enter into advance purchase agreements with pharmaceutical manufacturers to guarantee supply of these drugs [sic] and those countries that lacked the financial means to do so." When Indonesia attempted to procure the relevant supplies, it "confronted a queue, even though samples provided by Indonesia to the WHO had been used to make the vaccines and the country was recording the highest number of human-related H5N1 deaths."[28]

Moreover, the Advance Purchase Agreements that many industrialized countries, such as Canada, Switzerland, the US, New Zealand and more than half of EU member states, have in place with vaccine manufactures lie dormant until they become active and enforceable through a predetermined event, such as a phase six pandemic alert being declared by the WHO. Those agreements have a broad scope and may apply to novel influenza types that have not previously emerged. With limited production capacities in the time frame in which a pandemic can play out, they can effectively prevent third parties from obtaining vaccines as industry must fulfill its contractual obligations first. In addition to the unit costs once the agreement is triggered, contracting states are required to pay an annual fee to the vaccine industry in order for the agreement to be maintained.[29] Accordingly, "it may still be unrealistic to expect states with limited health care budgets [...] to spend a large proportion of that budget on maintaining a contract that guarantees vaccines for a pandemic which may not occur in the next 5, 10, or even 20 years."[30]

Finally, large disparities exist in terms of technological capacities within the GISN/GISRS network, which is comprised of four different types of institutions. National Influenza Centers are to collect and analyze relevant specimens and ship them to WHO Collaborating Centers and, since 2004, the H5 Reference Laboratories. The former "support outbreak investigation, conduct comprehensive virus analyses, and select and develop candidate influenza vaccine viruses with pandemic potential."[31] The latter conduct laboratory diagnosis of human infections in order to assess the risk (or existence) of a pandemic threat. All Collaborating Centers also double as H5 Reference Laboratories. Finally, Essential Regulatory Laboratories, which until recently were merely an informal component of the network, are linked to national regulatory agencies and develop, regulate and standardize influenza vaccines. Six Collaborating Centers, which serve as hubs within the GISN/GISRS network, presently exist worldwide: one each in Australia, China, Japan and the UK and two in the US. The four Essential

Regulatory Laboratories are located in Australia, Japan, the UK and the US. Those H5 Reference Laboratories that do not form part of Collaborating Centers are found in Cambodia, Hong Kong, Egypt, France, India and Russia. Besides the WHO-designated institutions, the US military operates a global network for biomedical research, which also collects samples of influenza viruses yet does not systematically share them with GISN.[32] This includes the Naval Medical Research Unit 2 in Jakarta, which became the subject of heated controversies in 2007 and was ultimately shut down by Indonesian authorities due to concerns that the shipping of samples through this parallel structure would undercut their attempts to restrict sample exports.[33]

Negotiations

First rumors of Indonesia having stopped cooperation with GISN emerged in December 2006. In the face of the pandemic threat from H5N1, the WHO's Secretariat already noted a month later that "[t]he sharing of H5N1 viruses and sequence data for vaccine research and development is [...] crucial for the protection of public health in all countries and is a collective responsibility. At the same time, however, the countries most severely affected by avian outbreaks and human cases are largely located in the developing world, and few have the capacity to manufacture pandemic vaccines."[34] The WHO Executive Board went on to recommend that sample sharing continue while noting that "[t]he technological benefits of participation [in GISN], including improved access to vaccines, should be available to all countries."[35]

Pressure was building up. Two months later, more than two dozen developing countries adopted the Jakarta Declaration on Responsible Practices for Sharing Avian Influenza Viruses and Resulting Benefits in order to "explore the modalities of a framework that strongly emphasizes the need for developing countries to share in the benefits resulting from the open and timely and equitable sharing and dissemination of information, data and biological specimens related to influenza."[36] The World Health Assembly took up the matter at its annual meeting in May 2007. The US proposed improving access to vaccines through stockpiling, donations, financial and technical support while emphasizing the necessity for continuing the unconditional sharing of virus specimens with GISN.[37] Developing countries asserted that the benefits arising from the utilization of influenza viruses with pandemic potential should be shared fairly and equitably and that access should be conditional on Prior Informed Consent.[38] Where benefit-sharing fails, provider countries would thus have the right to deny access to virus specimens collected within their territories. Price barriers and the hindrance of generic producers to manufacture patented vaccines were other causes of concern.[39]

The World Health Assembly resolved on a two-track approach. An Interdisciplinary Working Group would deal with "technical" aspects: a revision of the terms of reference for GISN institutions, some of which had been accused of exploiting their role in the WHO network for commercial ends and the creation of draft standard terms (i.e., an SMTA) to govern the sharing of samples from WHO member states with GISN and from GISN to third parties, such as vaccine manufacturers. An Intergovernmental Meeting (IGM) would deal with the "political" side by looking into novel mechanisms for fair and equitable benefit-sharing, with special hindsight to the interests of developing countries.[40] The Interdisciplinary Working Group only met a single time, in the summer of 2007 and the IGM process would continue until May 2009.

Access regulation and benefit-sharing

From 2007 to 2008, the chief political challenge was whether or not to link access to virus specimens with benefit-sharing: could each be addressed separately? Or would they be conditional on each other and addressed within a single, new regime? The US had previously ratcheted-up its voluntary contributions to global pandemic preparedness. Yet, as a US diplomat put it, "[w]e have a desire to be involved in both conversations" but "[w]hat we object to is linking them."[41] This linkage was precisely the objective of developing countries, with Indonesia in the lead. Claiming that viruses fell within the scope of the CBD, the right of provider countries to require Prior Informed Consent would secure the fair and equitable sharing of vaccines and other technology for viruses used for purposes other than "non-commercial risk assessment and response." Prior Informed Consent should also be required for the sharing of genetic sequence data through public databases.[42]

The African Group was the first to propose a two-tiered structure for ABS from viral materials. The transfer of (pandemic and seasonal) influenza viruses to GISN institutions would take place under different conditions than transfers from GISN to non-GISN institutions.[43] This would simultaneously allow for the rapid sharing of viral materials with WHO-designated laboratories and for the imposition of more stringent conditions on those users that subsequently generate monetary and health-related benefits. If non-GISN institutions would not comply with the terms and conditions under which transfers take place, Thailand opined, WHO members should have the right to cease the sharing of viruses.[44]

In the view of some WHO members, the right of governments to regulate access to virus specimens was put in doubt by the new International Health Regulations that had entered into force in 2007. The Regulations are a legally binding international treaty that seeks to "prevent, protect against, control and provide a public health response to the international spread of disease."[45] They contain certain reporting obligations: in the case of an "unexpected or unusual public health event," member states "shall provide to WHO all relevant public health information."[46] They also have the duty to report any potential public health emergency of international concern and to "communicate to WHO timely, accurate and sufficiently detailed public health information available to it on the notified event."[47] Does this entail a legally binding obligation to share virus samples? In the eyes of the EU, the US and other industrialized countries, the answer was "yes."[48] Requiring Prior Informed Consent or the conclusion of benefit-sharing arrangements before samples are being shared would thus be a breach of international law. This interpretation of the Regulations hinges on how broadly one interprets the term "information" and, accordingly, whether genetic sequence data or even physical specimen are included within their scope. Yet, even if the Regulations oblige member states to unconditionally share virus specimens or genetic sequence data with WHO, they are silent on what happens with the materials subsequently. Third parties outside of GISN could thus still be obliged to share the benefits they generate from the materials they receive.

The November 2007 IGM meeting being inconclusive, an Open-ended Working Group convened prior to the April 2008 World Health Summit and called for the development of a chair's text to speed up negotiations.[49] The IGM adopted this text as the basis for further negotiations in November 2008, with the "sharing of H5N1 and other influenza viruses with human pandemic potential" and the sharing of resulting benefits (antivirals and vaccines as well as through technology transfer and the building-up of vaccine production capacities in developing countries) henceforth being treated as equally important.[50]

This was a major breakthrough in the negotiations. It resulted from joint action by the US and Indonesia that moved the process from "the never-ending debate over voluntary vs. mandatory sharing of samples/benefits to commitments to share samples and to share benefits."[51] Bilateral US-Indonesian talks had commenced several months before the November 2008 meeting as, in the US view, IGM negotiators "would take note if the country which started the debate and the country most critical of Indonesia's position agreed on how to end the impasse."[52] The US concern was that Brazil, India and a few others attempted to use the negotiations as part of a strategy to "undercut" intellectual property rights.[53] From the side of developing countries, concerns remained that, despite the linkage between access and benefit-sharing, the actual commitments of industrialized countries were still far from concrete.[54]

Regulating nodes and vectors

To enhance both transparency of and control over the flow of PIP biological materials, developing countries had been pressing for a traceability mechanism and a mandatory SMTA governing the transfer of viral samples. The former issue was easily resolved. Already in November 2007, the IGM had requested the development of a mechanism to trace the flow of materials into and out of GISN. An interim mechanism dealing exclusively with H5N1 viruses went live in January 2008, with every such specimen circulating through GISN being assigned a unique identifier. A public database was installed containing information on their location, recipients and the analyses carried out on the materials themselves and derived seed viruses for vaccine production.

The SMTA (in other words the "terms and conditions" the World Health Assembly's resolution 60.28 had already foreseen in 2007) was more tricky. As a private contract between providers and users, it might contain limitations on the ways in which shared materials could be utilized, including restrictions on the claiming of intellectual property. The November 2008 IGM meeting agreed to use an SMTA for governing the transfers of viral specimens into and out of GISN, with outside institutions being barred from onward transfers outside of an SMTA, yet the devil was in the details. As the SMTA would contain the core obligations in regards to benefit-sharing, it remained the most contentious item on the negotiating table until 2010.

Beyond regulating the flows of PIP biological materials into, through and out of the GISN network, the WHO and its member states can directly control its nodes. The National Influenza Centers, H5 Reference Laboratories, Collaborating Centers and Essential Regulatory Laboratories each operate under their own terms of reference. Those are largely technical in nature yet, in principle, can also proscribe the ways in which GISN institutions shall (and shall not) utilize viral materials they receive or pass on. National Influenza Centers are the entry point for specimens into GISN. Their terms of reference directly touch upon the rights and obligations their respective governments have in regards to the sharing of viruses and genetic sequence data, such as the degree to which they may control their utilization after sharing with GISN has taken place. The terms of reference for WHO Collaborating Centers, the biggest fish in the GISN-pond, touches on the discretion they enjoy in both using and forwarding the materials they have themselves received: should Collaborating Centers be allowed to claim intellectual property, or should they be allowed to forward candidate vaccine viruses to manufacturers without the consent of the material's original provider? Should they be able to publish genetic sequence data in public databases without the provider's consent? And what should happen if they breach their terms of reference?

The Interdisciplinary Working Group the World Health Assembly had installed with resolution 60.28 was to deal with precisely those issues. Meeting in August 2007, the broad definition of the Group's mandate led to discussions also covering benefit-sharing and intellectual property rights. This ensured that deliberations would not remain purely "technical" but become politically charged. The WHO African Region, for instance, proposed that H5 Reference Laboratories, including the Collaborating Centers, would "[a]ctively seek approval" of provider countries before sharing H5 materials with other entities, thus introducing a Prior Informed Consent requirement through the backdoor.[55]

The question of who should enjoy which rights and obligations in regards to the utilization and transfer of virus specimens remained unsolvable at both the technical and the political level. The Working Group failed to produce an agreed-upon outcome. The subsequent November 2007 IGM session equally failed to make progress on the issue. Only in December 2008 did the IGM agree not on the terms of reference themselves, but rather on a set of guiding principles for the development of such terms and references. Those included obliging GISN institutions to use the new Influenza Virus Traceability Mechanism for monitoring flows and to use and transfer virus specimens in line with an eventual SMTA.[56] Two principles that did not find consensus at first yet were incorporated into the Framework in 2011 were the sharing of genetic sequence data with relevant databases and the obligation to give originating laboratories "due credit and recognition" for the PIP materials they provide.

Based on the guiding principles, the IGM adopted several core terms of reference at its final session in May 2009. An important component was that WHO Collaborating Centers would share genetic sequence data within GISN, yet uploads to publicly available databases would require the consent of the laboratories originally providing the physical specimens, in other words, the provider country. The singular issue that remained unresolved for the time being was whether Collaborating Centers would be obliged to acquire the provider's approval before passing on specimens for use in commercial applications such as vaccine production.

The end of the IGM

When the IGM negotiations terminated in May 2009, delegates had found consensus on large parts of the future Pandemic Influenza Preparedness Framework, including most principles, technical definitions and Terms of Reference, scope, traceability and the benefit-sharing language of the Framework itself, though not the SMTA. As the SMTA would give practical effect to WHO members' general and unspecific commitment towards the sharing of benefits, it was at the core of the negotiation process. Should it apply only the transfer of viral materials or also their use? Should it cover transfers and/or uses by non-GISN institutions that receive materials from the network? Would it be "entered into under the Framework" or merely "interpreted in accordance" with the Framework?[57] Finally, should patent claims on shared materials entail an obligation to grant to WHO a royalty-free, non-exclusive and transferable license that would subsequently be transferred to developing countries?

Two years of talks had brought little, if any, progress on the matter. Some of the Framework's principles remained equally disputed. Developing countries pushed for the explicit recognition of state sovereignty over biological resources, and the application of this principle to viruses (and their components and derivatives) moved through the system. A reference to the Doha Declaration on the TRIPS agreement and public health, which had reaffirmed the flexibilities international patent law grants in domestic implementation, equally went against industrialized countries' agenda of keeping intellectual property out of the negotiations. Finally, the issue of whether GISN institutions should be allowed to use circulating viral materials for purposes other than diagnostics, risk assessment, selection of candidate vaccine viruses and vaccine development without prior permission remained contentious.[58] Those were the issues that had prompted negotiations on an ABS regime for viral materials in the first place, yet negotiations had so far accomplished little in overcoming the "breakdown of trust" that the draft Framework noted in a non-agreed part. The US and others were perceived as attempting to "scuttle" the negotiations.[59] Industrialized countries were mistrustful of Brazil and India's alleged hidden agenda in regards to intellectual property rights.

In May 2009, the World Health Assembly instructed the WHO Director General to "work with Member States to take forward the agreed parts" of the Framework and "facilitate a transparent process to finalize the remaining elements."[60]

The rest of the year was dedicated to informal consultations on the sticking points: the SMTA, intellectual property rights and the relationship to the CBD and its principle of state sovereignty.[61] WHO Director General Margaret Chan had to walk a fine line between those countries concerned that a broad SMTA would "slow down the transfer of vaccines and related materials at precisely the time when speed is most needed," and those arguing that "benefit-sharing commitments will be avoidable and developing countries will be left without affordable access to needed medication" under a narrow one.[62]

The resulting Chan proposal was based on a single SMTA for transfers within GISN, with recipient manufacturers having to have in place, or be in the process of developing, benefit-sharing arrangements with the WHO. Where patents are claimed, right holders "should" grant a non-exclusive, free and transferable license to the WHO.[63] The reaction of developing countries was less than enthusiastic. Brazil, India, Indonesia, Nigeria and others emphasized that the outcome of the process would need to deliver a "sustainable solution" to the vaccine problem instead of *ad hoc* measures such as donations, which, as experience had shown, were unlikely to be forthcoming during a global health emergency.[64] Formal negotiations only resumed when the WHO Executive Board decided to convene an Open-ended Working Group prior to the 2010 World Health Assembly. At the same time, developments under the CBD were picking up pace.

The Nagoya connection

The negotiations on the Nagoya Protocol were entering their final stretch when, in early 2009, a coalition of developing countries began pushing for including pathogens within its scope. At the 7th meeting of the CBD's ABSWG, the Like-Minded Group of Megadiverse Countries urged WHO negotiators to recognize the applicability of the CBD, recalling that resolution 60.28 had reaffirmed state sovereignty over biological resources. The WHO negotiations were thus not to prejudge the outcome of the CBD process.[65] The African Group also voiced concerns over the "sectoralization" of ABS governance as more and more types of genetic resources would fall under specialized instruments outside of the CBD regime.[66]

The Like-minded Group of Megadiverse Countries was instrumental in bringing pathogens onto the agenda of ABSWG 7 in what was possibly an attempt at creating "leverage for striking deals in future meetings."[67] Among the members of the group were those developing countries, Brazil, India and Indonesia, which had taken on leading roles within the WHO negotiations. The issue remained unresolved up to and including the October 2010 Nagoya Summit. Due to the expected inconsistencies with both intellectual property rights and pandemic preparedness and response, industrialized countries considered the issue a "potential deal-breaker."[68]

The final Nagoya Protocol does not explicitly cover viral materials. As it applies to all genetic resources within the scope of article 15 CBD, the relevant question is, then, whether such materials constitute genetic resources under the CBD. This is a complex legal discussion that mainly revolves around the question of whether viruses are *human* genetic resources and are thus excluded from the CBD's scope. Neither the Protocol nor the eventual Framework gives a satisfactory answer to this question. However, the Protocol's article 4.4 creates an exemption for specialized ABS regimes. Where those are "consistent with" and do not "run counter to" the objective of fair and equitable benefit-sharing, the specialized regime applies whereas the Protocol does not. The Protocol's article 8.b further invites parties to "take into consideration the need for expeditious access to genetic resources" and equally expeditious benefit-sharing in cases of health emergencies, and the preambular text, "[m]indful of the International Health Regulations," emphasizes "the importance of ensuring access to human pathogens for public health preparedness and response purposes." As I discuss below, this merely avoids (or postpones) the difficult political decision of whether or not governments have the right to regulate access to viruses, or not.

Deliberate ambiguity

The conclusion of the Nagoya Protocol upped the ante for WHO negotiators: without a specialized regime for viral materials, developing countries could claim the full applicability of the Protocol during the next pandemic emergency. Meeting two months after the Nagoya summit, the Open-ended Working Group attempted to clarify the relationship between the Protocol and the future Framework. The negotiating text coming out of the December 2010 meeting contained some important changes when compared to the 2009 IGM text. Delegates dissented on whether to recognize Nagoya Protocol articles 4.4 and 8.b, thus making the Framework into a specialized ABS regime, or whether to affirm the exclusion of viral materials from the scope of both the CBD and the Protocol.[69] The latter would have effectively eliminated institutional overlap, as the Framework would simply deal with an entirely different subject matter. The former would have acknowledged the applicability of the Protocol and the CBD, with the article 4.4 exemption only being active as long as benefits, such as vaccines, are being shared fairly and equitably.

Delegates managed to square the circle at the final negotiating session before the Framework's adoption by the World Health Assembly in May 2011. At the final meeting of the Open-ended Working Group, they agreed to recognize state sovereignty over genetic resources as one of the Framework's principles while simultaneously leaving open the question of whether or not this principle actually *applies* to viruses, their components and derivatives. While the CBD understands "genetic resources" broadly as "any material of plant, animal, microbial or other origin containing functional units of heredity" that is "of actual or potential value,"[70] the hereditary information viruses contain is not necessarily "functional" insofar as it only reproduces in host organisms; moreover, understanding viruses as *human* genetic resources would exclude them from the scope of the CBD in line with a 1995 decision by the CBD's Conference of the Parties. The extent to which the state sovereignty principle thus practically matters for purposes of the Pandemic Influenza Preparedness Framework is thus open to quite different interpretations.[71] This deliberate ambiguity made it possible for each side to avoid language that would prove unacceptable to the other. At the final negotiating session, parties proposed that the World Health Assembly, in its resolution for the adoption of the Framework, would clarify that the Framework is either a specialized instrument in line with Article 4.4 Nagoya Protocol or simply an ABS instrument for viral materials.[72] Yet neither the eventual resolution nor the Framework itself would end up clarifying the relationship with the CBD and the Protocol.

The SMTA 2

Besides the creative solution for the conflict over the Framework's relationship with the CBD, the months before its adoption by the World Health Assembly saw a development of arguably larger practical significance. A December 2010 proposal by Brazil, India and Indonesia resulted in one of the defining features of the Framework: the use of two separate SMTAs governing, respectively, transfers to and within GISN and onward transfers to third parties, including vaccine manufacturers. The original proposal would have prevented GISN institutions, operating under SMTA 1, from seeking intellectual property protection for viral materials and associated inventions. While this restriction would not apply to non-GISN institutions receiving materials under SMTA 2, they would be obliged to grant a royalty-free, non-exclusive and transferable license to the WHO.

The use of two SMTAs in the final Pandemic Influenza Preparedness Framework addressed a major concern of developing countries: the unregulated use of PIP biological materials by institutions outside of the GISN network, now renamed the Global Influenza Surveillance and Response System (GISRS). The text adopted by the World Health Assembly in May 2011, however, was less ambitious than foreseen by Brazil, India, Indonesia and others. Under SMTA 1, neither providers (the National Influenza Centers) nor receiving GISN institutions (Collaborating Centers and H5 Reference Laboratories) "should" seek intellectual property rights on shared materials, a provision that falls short of a categorical ban. Under SMTA 2, recipients must choose among several options for the sharing of resulting benefits. This includes the option of granting royalty-free, transferable and non-exclusive licenses to WHO, yet third parties may also opt for other forms of benefit-sharing, including vaccine donations or their sale at preferential prices. Both SMTAs mandate the use of the Influenza Virus Traceability Mechanism to monitor the flow of viral materials into, within and out of GISRS, and no transfers may take place outside of either SMTA 1 or 2, as applicable.

Explaining institutional outcomes

The Pandemic Influenza Preparedness Framework and its supporting components (the Traceability Mechanism, Terms of Reference for GISRS institutions, the Advisory Mechanism and the two SMTAs) differs in important ways from the earlier GISN network. It requires mandatory benefit-sharing; provides for the monitoring of viral flows into and out of GISRS; acknowledges state sovereignty over biological resources without clarifying whether or not this entails viral materials; encourages GISRS institutions to not seek intellectual property protection on the materials and related inventions; and allows commercial users to voluntarily share their intellectual property with the WHO. The Nagoya Protocol, finally, attempts to establish a division of labor with the Framework by implicitly recognizing it as a specialized ABS regime.

Virus-sharing for pandemic preparedness and response, which had been a technical exercise for decades, has thus come to include a number of new and heavily politicized issues. How to explain those changes? As I argue below, the interplay management hypothesis initially appears as a strong contender yet suffers from serious shortfalls once we take a closer look. For regime shifting and situation structure, there are important overlaps: the former assumes that changing the predominant regulatory approach to viral materials, namely private property rights, was the chief factor driving developing countries' bargaining behavior. Yet switching the applicable regime is equally a lever for obtaining more gains from international cooperation. Regime shifting has higher explanatory power than in the other three cases discussed in this book and is to a large extent observationally equivalent to situation structure.

Interplay management

The interplay-management explanation suggests that bargaining aimed at improving pandemic influenza response through better access to vaccines while ensuring a consistent relationship between the Framework on one hand, the CBD, the patent regime and the International Health Regulations on the other. To a certain extent, negotiators attempted to manage the problematic relationship with the CBD in order to ensure smooth production, supply and distribution of vaccines. However, the only reason this problematic relationship existed in the first place was the argument by Indonesia and its supporters that providers of viral materials enjoy sovereign rights and can thus exert control over the materials' subsequent utilization. In other words: the problem requiring interplay management only emerged because some actors deliberately created it.

But what is this problem, precisely? The entire debate revolved around whether or not viral materials constitute "genetic resources" in line with the CBD. If that is so, national governments have the right to determine access under CBD article 15.1. For viral materials accessed in this manner, in turn, the Nagoya Protocol would apply in full. But the legal status of viral materials is left open under both the Framework and the Protocol. The former acknowledges state sovereignty in its preambular text yet does not mention whether state sovereignty extends to the materials falling under its operative text. The Framework also does not speak of "genetic resources" but rather adopts the terminology of "Pandemic Influenza Preparedness biological materials." The phrase "biological materials" does not appear in the CBD, which only speaks of "biological resources," "genetic material" and "genetic resources."

From the side of the Protocol, the question is whether article 4.4 can be understood as an instance of interplay management that establishes a division of labor in respect to the Framework: as long as specialized ABS regimes are "consistent with" and "do not run counter to" the benefit-sharing objective, the specialized regime applies instead of the Protocol. Yet there is an important problem hidden here. The article 4.4 exemption exclusively applies to *genetic resources*, more precisely: those genetic resources that fall within the scope of the Protocol, which itself covers only those genetic resources within the scope of article 15 CBD. The implication is: if the article 4.4 exemption applies to viral materials, those materials are indirectly acknowledged as genetic resources within the scope of article 15. The Nagoya Protocol would not apply as long as the benefits arising from the utilization of viral materials are being shared fairly and equitably. If benefits are *not* thought to be shared in this manner, for instance because insufficient vaccine manufacturers have concluded SMTA 2s with the WHO, the Protocol applies by default. Contracting parties would thus be obliged to apply the articles 15, 17 and 18 compliance measures to domestic users of viral materials. What is more, with viral materials falling within the scope of article 15, national governments have the right to determine access and require Prior Informed Consent. Of course, none of those problems exists if viral materials are not considered as genetic resource or are considered as *human* genetic resources explicitly excluded from the scope of the CBD. As mentioned, the drafters of the Framework and the Protocol managed to avoid precise definitions to paper over the political conflicts.

Thus, the article 4.4 construct does not resolve anything. Instead, it enables a repeat of the entire debate over the status of viral materials once the next influenza pandemic comes around: if vaccines and other benefits are being shared with developing countries in line with the Framework, the legal argument could be that this triggers the article 4.4 exemption. Then again, if benefits are shared, there is little reason for developing countries to invoke state sovereignty over viral materials in the first place. Conversely, if benefits are *not* being shared, developing countries will claim, on the basis of article 4.4 as well as article 8(b), that they have the right to determine access to viruses collected within their respective territories. In this context, it should be noted that, more than five years after the adoption of the Framework, SMTA 2s are in place with a mere four manufacturers, two of which are from China and India, respectively.[73] In the meantime viral materials continue to be shared through GISRS under an interim process that obliges recipients to enter into an SMTA 2 in the future.

The Framework equally fails to clarify the extent to which the International Health Regulations apply to the sharing of PIP biological materials. Do they contain an obligation for WHO members to unconditionally and rapidly share physical specimens, genetic sequence data or both? Or do those fall outside the scope of the "accurate and sufficiently detailed public health information," including laboratory results such as from National Influenza Centers, under article 6.2? If they do, the notion of national sovereignty over viral materials, and the concomitant right of nation states to require Prior Informed Consent before access takes place, is moot.

The central dispute of the entire negotiation process thus remains unresolved. Things look better for the Framework's operational provisions on access to viral materials and the sharing of benefits. Unlike in the case of the Seed Treaty, SMTAs 1 and 2 manage to avoid complications in regards to patenting. SMTA 1 only contains the provision that providers and recipients "should" not seek patent protection, whereas SMTA 2 allows recipients to voluntarily grant royalty-free, non-exclusive and transferable licenses to WHO as one among several options for benefit-sharing. As the TRIPS agreement unconditionally requires WTO members to grant patents on vaccines and related technologies (with the exception of diagnostic, therapeutic and surgical methods),[74] limitations on patent claims would arguably be of a more serious nature than in the case of the Seed Treaty, where WTO members may choose to exclude plants from patentability.

Regime shifting

The Pandemic Influenza Preparedness Framework is a strong case in point for regime shifting. Institutional outcomes are congruent with what the hypothesis suggests: the utilization of pandemic influenza viruses is subject to a new regulatory approach, with the possible application of state sovereignty challenging the right of pharmaceutical and vaccine manufacturers to unconditionally access and utilize viral materials exclusively for their own commercial ends. With the ambiguous relationship of the CBD, the Nagoya Protocol and the Framework in regards to viral materials, states reserve the option of claiming state sovereignty and requiring Prior Informed Consent. While the Framework and its SMTAs steer clear of infringing on patent law, the position of developing countries has been strengthened by the implicit assertion of ownership over those viruses that are passed through GISRS to third-party users.

This raises the question of observational equivalence: to what extent are the processes leading up to the PIP Framework explicable in terms of regime shifting (i.e., pushing back private property rights over vaccines and relevant technologies by asserting state sovereignty)? To what extent in terms of situation structure (i.e., using state sovereignty as a lever to gain enhanced access to vaccines and other benefits)? There is no proper way of adjudicating between the two hypotheses. Unlike in the case of Farmers' Rights, where concessions came early on and possibly outside of a larger package deal, developing countries made few concessions on viral sovereignty and those only in the very final stage of the negotiation process, when their insistence on the formal recognition of state sovereignty posed the risk of scuttling the entire negotiation process.

There are two other data points in favor of regime shifting as a factor driving the bargaining behavior of Indonesia and others: first, the Core Terms of Reference for WHO Collaborating Centers, some of which were at the heart of the 2006/2007 controversy. The new terms require the Centers to "appropriately acknowledge" National Influenza Centers providing viral materials, including in "presentations and publications."[75] While this does not touch on the question of illicit patent claims by the Centers, it enhances the status of provider countries in multilateral virus-sharing. Second, the requirement that recipients of materials under SMTA 1 "should" not claim intellectual property can be read both as an attempt to avoid conflicts with countries' obligation under international law to grant patents once certain conditions are met (interplay management) and the legally non-binding, yet symbolically relevant acknowledgment that the Centers should not exploit their public role within GISRS for the pursuit of private and commercial gains. Both the Core Terms of Reference and SMTA 1 can thus be understood as attempts to gear multilateral virus-sharing more strongly towards the public interest of WHO members rather than the interests of commercial vaccine manufacturers and related industries. Both are arguably unrelated to the distribution of material gains from cooperation and thus favor an explanation in terms of regime shifting over one in terms of situation structure.

Regardless of the strength of the hypothesis, the resulting regime shift is only partial. The inclusion of pathogens within the scope of the Nagoya Protocol ultimately failed, and the Framework only acknowledges sovereign rights over viral materials in an, at best, implicit manner. Thus, we need to look not only into the question of why developing countries *attempted* to regime-shift, but also at the likely reasons they were partially unsuccessful in doing so.

Situation structure

Situation structure implies that the negotiations on the Framework neither had the goal of resolving tensions between international regimes nor the goal of shifting the applicable international regulatory frameworks, but rather to compel industrialized countries to ensure the effective and equitable supply of developing countries with vaccines and other technologies during a pandemic emergency. The question is then: to what extent was the struggle over the application of state sovereignty to viruses simply an attempt to leverage access regulation for the extraction of benefits (situation structure), and to what extent did it matter whether the applicable regulatory approach would be changed (regime shifting)? Considering that developing countries never gave up on viral sovereignty and pressed hard for its inclusion in the Framework, resulting in its ultimately ambiguous incorporation into the preambular text as well as the problematic article 4.4 of the Nagoya Protocol, this question cannot be answered with certainty.

However, virtually all industrialized countries strictly preferred the pre-2007 status quo and were only willing to move forward on the Framework due to fears of a possible domino effect, with more and more developing countries ceasing to share samples. The former were unwilling to accept any outcome that would imply lengthy access negotiations in case of a pandemic emergency.[76] The counterfactual, to which I turn below, can answer the question of who had more to lose from negotiation failure, either on the Framework or on the Nagoya Protocol. However, assessing the congruence between institutional outcomes and interests weighted by interdependence is difficult. The Framework does impose costs on vaccine manufacturers through both the SMTA 2 and the Partnership Contribution, requiring recipients of viral materials to collectively finance half of GISRS's annual operating costs. Compared to the way in which viruses were shared multilaterally before the Indonesian controversy broke, the Framework offers only disadvantages to user countries. At the same time, non-users have improved their position by having vaccine manufacturers enter into SMTA 2s in the future, thus leading to foreseeable and effective benefit-sharing instead of voluntary philanthropy. Another question is whether the costs of benefit-sharing under SMTA 2 are so high as to constitute a significant drawback for manufacturers. Some authors are skeptical of the depth of the benefit-sharing obligations, arguing that the Framework "secured the norm of virus sharing while providing only weak benefit sharing in return."[77]

The only major data point suggesting that the better placement of user countries within the situation structure decisively shaped institutional outcomes is the exclusion of genetic sequence data from the definition of "PIP biological materials": their utilization thus does not entail an obligation to share benefits.[78] The transfer of sequence data from GISRS institutions to manufacturers is largely unregulated under the Framework: it is covered neither by the Traceability Mechanism nor by SMTA 1 or 2. The increasing adoption by the vaccine industry of technologies for the creation of synthetic viruses based on sequence data and not physical specimens implies that benefit-sharing under SMTA 2 will slowly whittle down unless there are legal changes to the Framework.[79] Considering that the legal status of sequence data is ambiguous under the CBD as well, this development could make the question of whether or not viruses are "genetic resources" irrelevant in the long run.

The club counterfactual

Pandemic influenza viruses are not "provided," they emerge at largely random locations in the world as a result of genetic mutation. While the case can be made that the endemism of viruses with pandemic potential is higher in some regions than in others, the precise origins of pandemics can be a matter of dispute: the "Spanish" flu probably did not arise in Spain but rather in France or Austria.[80] Yet considering that pandemic influenza may, in principle, originate anywhere in the world, the distinction between "users" and "non-users" is more helpful than between "users" and "providers." When the Indonesian controversy broke, Northern governments wanted things to remain the way they had been for several decades, whereas (some) Southern governments, the most vocal being Indonesia, wanted better access to pandemic influenza vaccines and related technology than was being made available under the status quo. Under the Pandemic Influenza Preparedness Framework, both sides made compromises. What would have been the consequence if club cooperation had allowed them to forego compromises and instead create rules closely aligned with the interests of club members?

On a very general level, the probable result would have been a significant danger to global public health, although users would have been better off: assuming that users and non-users have an equal risk of being the origin of an influenza pandemic, once an outbreak hits, users would have both the virus and the resultant vaccines, whereas non-users would have the virus but no possibility of rapidly rolling out sufficient amounts of vaccines. This does not mean that minilateral forms of cooperation are ineffective as such: the US Naval Medical Research Unit Two, with branches in several Southeast Asian countries, participates within the multilateral GISN/GISRS regime but also directly forwards relevant samples to the US Center for Disease Control. With numerous infectious diseases being endemic in the region, sample transfer based on bilateral agreements allows, in principle, for US vaccine manufacturers to procure the relevant materials even in the absence of multilateral cooperation. In response to the H1N1 pandemic having emerged in North America, the governments of Canada, Mexico and the US developed a trilateral framework, the North American Plan for Animal and Pandemic Influenza, which includes obligations on virus sharing.[81] The European Influenza Surveillance Network is a similar structure at the regional level, which runs parallel to GISN/GISRS.

The second-order counterfactual for club cooperation in combination with negotiation failure is very different from the counterfactual for negotiation success. Failure would probably have led users to prioritize health security over equitable access to vaccines, thus aggravating the problems of non-users. In the case of the successful conclusion of the Pandemic Influenza Preparedness Framework without the participation of users, the latter might have had incentives to offer concessions, in terms of improved access to vaccines, in order to join the club, yet until that time, non-users would initially have failed to secure their original policy objective, which was precisely improved access to vaccines. For non-users, the second-order counterfactuals are this: if they had entered into a club that requires the sharing of vaccines and related technologies among its members, non-members (users, that is) would have had no reason to finalize the negotiations on the Framework, seeing that it obliges their domestic vaccine manufacturers to enter into SMTA 2s on behalf of other countries. Club cooperation among non-users under the second-order condition of negotiation failure, conversely, would have essentially left them where they began: in a situation where access to vaccines and other technologies is contingent on "ad hoc charitable donations."[82]

The thought experiment shows that club cooperation was not a viable option for either group of actors, yet it also shows asymmetrical interdependence at work: user clubs would have been better off than non-user clubs, also considering the former's better public health infrastructure. Still, considering the risks at stake, it is not surprising that nobody chose to go down this path.

Notes

1 Price-Smith (2009).
2 Kamradt-Scott (2012), 90.
3 Hameiri (2014).
4 Avian Influenza Action Group (2008); IPW (2008a).
5 Gostin (2014), 373–374.
6 Gostin (2014), 361–362.

7 Gostin (2014), 361.
8 Lavanchy and Gavinio (2001), 10.
9 Kamradt-Scott (2015), 126–127.
10 Hammond (2009).
11 Irwin (2010).
12 07Jakarta264.
13 07Jakarta310; see also Sedyaningsih et al. (2008).
14 Sedyaningsih et al. (2008), 485; Global Health Watch (2008), 233.
15 Fidler (2007).
16 Vezzani (2010); see Section 7.5 for greater detail.
17 Smallman (2013), 24–26.
18 07Jakarta310.
19 07Brussels1459.
20 Quoted in Hammond (2009).
21 Doherty (2013), 70–71.
22 Price-Smith (2009), 82–83.
23 PhRMA (2013), 45; see also Stöhr and Esveld (2004), 2196.
24 NAS (2009), 109.
25 Abbott and Dukes (2009), 116–162.
26 Price-Smith (2009).
27 Elbe (2011), 849, italics i.o.
28 Kamradt-Scott (2015), 158–159.
29 Turner (2015).
30 Turner (2015), 5.
31 A62/5 Add. 1, 40.
32 Hammond (2009).
33 08Jakarta806.
34 EB120/15, para 12.
35 A60/Inf.Doc/1.
36 Jakarta Declaration, paras 1 and 5.
37 WHA60/2007/REC/3, 13.
38 WHA60/2007/REC/3, 15.
39 WHA60/2007/REC/3, 21–28.
40 WHA Resolution 60.28, paras 5 and 7.
41 Quoted in IPW (2008a).
42 A/PIP/IGM/5, Annex.
43 A/PIP/IGM/7, 11.
44 A/PIP/IGM/8.
45 International Health Regulations, Article 2.
46 International Health Regulations, Article 7.
47 International Health Regulations, Article 6.2.
48 SUNS (2007); see also EB122/2008/REC/2, 42.
49 A/PIP/IGM/WG/5, paras 10 and 11.
50 A/PIP/IGM/WG/6, 5 and 13.
51 08Geneva1112.
52 08Jakarta1887.
53 08Geneva1112.
54 IPW (2008a).
55 A/PIP/IGM/7, 23.
56 EB124/4 Add.1, 8.
57 A62/5 Add.1, 31.
58 A62/5 Add.1, 16.
59 TWN (2009), 2.

60 WHA Resolution 62.10.
61 EB126/4.
62 IPW (2009), 2.
63 Ibid.
64 Cited in TWN (2010), 3.
65 UNEP/CBD/WG-ABS/7/8, 16.
66 ENB (2009).
67 ENB (2009), 13.
68 Buck and Hamilton (2011), 58.
69 EB128/4, 9.
70 CBD, Article 2.
71 Abbott (2010).
72 A64/8, 52.
73 see http://www.who.int/influenza/pip/benefit_sharing/smta2_signed/en/, accessed 22 September 2016.
74 TRIPS article 27.2.
75 Pandemic Influenza Preparedness Framework, Annex 5, A(8) and C(1).
76 Buck and Hamilton (2011), 58.
77 Gostin (2014), 377; see also Kamradt-Scott and Lee (2011).
78 Gostin et al. (2014).
79 The Advisory Mechanism is mandated to look into the handling of genetic sequence data. This is an ongoing process. See Pandemic Influenza Preparedness Framework, article 5.2.
80 Price-Smith (2009).
81 NAPAPI (2012), 29–30.
82 Gostin (2014), 373–374.

8 Marine genetic resources

The final case study of this book deals with a moving target: the ongoing process towards an international regime for the protection of marine biodiversity in Areas Beyond National Jurisdiction (ABNJ). This case is peculiar: it involves institutional layering in two originally separate issue areas: the conservation and sustainable use of biodiversity in international waters, and ABS from Marine Genetic Resources (MGR).The linkage between the two is what enables institutional layering in both issue areas, and the eventual conclusion of the ongoing negotiation process will transform the governance architectures for both. In the case of marine biodiversity in ABNJ, what is at stake is the centralization of the presently highly fragmented system of ocean governance that involves various sectoral and regional approaches yet lacks a central, coordinating institution. In the case of MGR, the question is how to implement benefit-sharing in an issue area for which it was originally not intended, that is, beyond the confines of national jurisdiction. The process towards a new international agreement encompassing both MGR and marine biodiversity in ABNJ commenced in 2003 yet only gained traction in 2011 due to three coinciding factors: the conclusion of the Nagoya Protocol in the previous year; political pressure to combat ocean degradation in the run-up to the 2012 Rio +20 summit and a broad coalition between the EU and the G77/China, which pushed for a broad package deal containing ABS and biodiversity protection as part of a single undertaking.

At the time of writing, a Preparatory Committee under the UN General Assembly is negotiating draft text for a legally binding instrument that will serve as input to a diplomatic conference, should the 2018 General Assembly decide to convene it. Thus, there are no institutional outcomes against which my hypotheses can be tested. I thus focus exclusively on the negotiation process. Here, the interplay management-hypothesis fares extremely well, as the careful insertion of the new international agreement into the existing institutional architecture for oceans governance and the avoidance of potentially disruptive institutional interactions has been a focal point of the entire negotiation process. Accordingly, regime shifting provides little explanatory value as several developing countries have already signaled their willingness to move away from the only relevant alternative regime, the common heritage of mankind approach to MGR, given that adequate arrangements for benefit-sharing will be agreed on in a pragmatic manner instead. Assessing the power of the situation-structural approach, finally, faces the problem that the ABS component of the new regime for marine biodiversity is but a small aspect of a significantly larger package. This sets the present case apart from the cases discussed in Chapters 5 to 7 where ABS-related questions were the focal, or even exclusive, area of negotiations. A full exploration of the underlying situation structure is beyond the scope of this book as it involves numerous and disparate elements such as high-seas fishery, marine ecosystem services, marine pollution from various types of point and diffuse sources, ocean acidification and a host of other issues. To simplify the analysis, I assess situation structure only insofar as it pertains to MGR. The consequence is that the hypothesis cannot adequately explain the proactive role of the EU in regards to ABS from MGR in the context of its broader alliance with the G77/China.

Marine biodiversity, blue biotechnology

The larger context in which negotiations on an international instrument in ABNJ play out is similar to that which gave rise to the Seed Treaty: on one hand, marine biodiversity is being lost at a rate that jeopardizes the sustainability of ocean ecosystems in the long term. On the other, commercial and scientific interest in the "blue gold" of MGR is rapidly accelerating, giving rise to calls for sharing the benefits arising from their utilization fairly and equitably. Both issues face a protracted legal problem: the oceans are contiguous, yet international law establishes different regimes for discrete areas. States possess full, sovereign rights within 12 nautical miles (nm) from their respective coastal baselines. The contiguous zone, within which states enjoy certain enforcement rights, stretches out another 12 nm. From the territorial seas up to 200 nm from the baseline is the Exclusive Economic Zone, where states have the sovereign right to exploit to resources under the sea, whereas its surface is already considered international waters. Beyond the Exclusive Economic Zones are international waters where no state may claim sovereignty. Still different rules apply for the states' continental shelf. In international waters, the applicable law also discriminates between the seabed, the ocean floor and its subsoil ("the Area") on one hand and the water columns above it on the other.

ABNJ, which include the high seas, are the "least known and least protected areas on Earth."[1] The oceans are facing threats on multiple fronts. Fishery, especially in its Illegal, Unreported and Unregulated (IUU) variety, is arguably the most drastic one, and one a whole battery of international treaties appears unable to resolve. A combination of improving technology, government subsidies and the inability to effectively regulate access to the scarce resource of global fish stocks is leading to overexploitation with potentially disastrous impacts on global food security.[2] Fishery is far from the only problem threatening marine ecosystems. Maritime traffic entails the routine discharge of oil, and invasive alien species may be transported into vulnerable ecosystems through ballast tanks. Pollution from land-based sources includes highly toxic chemicals that are transported over long distances. Hazardous wastes are being dumped on the high seas.

Oil spills regularly cause massive ecological crises. Anthropogenic emissions of carbon dioxide not only find their way into the atmosphere, but roughly half of them are sequestered by the oceans where they lead to acidification that, in turn, threatens marine ecosystems and especially coral reefs, among the most-biodiverse areas on the planet.[3] This "escalating crisis in marine ecosystems [...] is in large part a failure of governance."[4]

One major factor spurring the development of an international regime for biodiversity in ABNJ, the topic of this chapter, is the increasing institutional fragmentation of marine governance. The United Nations Convention on the Law of the Sea (UNCLOS) provides an overarching framework and general principles. Multilateral processes on Illegal, Unreported and Unregulated Fishing take place in the International Maritime Organization and the FAO, which adopted the Port State Measures Agreement in 2009. The International Seabed Authority (ISA), established by UNCLOS, addresses the environmental impacts of deep sea mining. The UN Fish Stocks Agreement is an international treaty implementing basic principles of UNCLOS in regards to straddling and migratory fish in ABNJ.[5] Ocean dumping and vessel-source pollution are covered, respectively, under the 1972 London Convention and the 1973/1978 MARPOL Convention. The CBD applies to "processes and activities" in ABNJ that are under the jurisdiction or control of contracting parties that, in the case of the high seas, refers to flag state jurisdiction.[6] The CBD's COP has also adopted several decisions on Marine Protected Areas and has put in place an Ad Hoc Working Group on Protected Areas. Aichi Target 11 foresees the creation of protected areas for 10% of coastal and marine ecosystems by 2020.[7] Since 1999, a high-level UN consultative process addresses the safety of navigation and the protection of vulnerable marine ecosystems. Both the 2002 World Summit on Sustainable Development and the 2012 UN Conference on Sustainable Development ("Rio +20") have adopted various targets relevant to ocean governance. Regional Seas Agreements covering topics from ocean dumping and pollution from land-based sources to international cooperation in the case of oil spills exist in the Mediterranean, the Baltic Sea, the North Sea, the Red Sea, the Indian Ocean and several other areas. Regional Fisheries Management Organizations are in place, *inter alia*, in the Indian Ocean, the Atlantic, the Antarctic and the South Pacific.[8]

This is only a snapshot. Interest differentials between those countries preferring conservation of high-seas biodiversity and those preferring exploitation explains the lack of a comprehensive, global framework for integrated ocean management.[9] A major bone of contention in the negotiations on an international regime for marine biodiversity in ABNJ is the question of whether there are governance gaps, implementation gaps or regulatory gaps within this fragmented governance architecture. One particular issue is the sectoral approach of existing instruments, each addressing different aspects of biodiversity protection but none taking a comprehensive, "ecosystem-based" approach.[10] For some delegations, "integrated management approaches were needed to bring current sectoral authorities and tools together."[11] The problem partially derives from UNCLOS itself, which emphasizes both regional and sectoral approaches to the protection of ecosystems in ABNJ.[12] Due to the special legal status of international waters, moreover, UNCLOS currently only allows the regulation of human activities through flag state jurisdiction instead of direct conservation targets for ecosystems themselves. This is a problem where harmful activities would originate from *outside* of protected areas.[13] The applicability of the CBD is equally doubtful. While it requires contracting parties to "establish a system of protected areas [...] to conserve biological diversity," this only applies to "processes and activities" in ABNJ under their jurisdiction; in the case of international waters, this is flag state jurisdiction.[14]

Discussions on ABS from MGR latch on to this negotiation process and contribute to further agenda overload. When UNCLOS was concluded in 1982, the biotechnological utilization of MGR was on nobody's radar. Only recently have utilization and concomitant patent claims taken off. While the oceans contain about 70% of the Earth's biosphere, the technology for accessing MGR has only recently become available. Unlike for terrestrial biodiversity, the sunk genetic treasures of the oceans are still largely unknown, yet they are potentially of significant scientific and commercial value. For instance, one study suggests that the "hit-rate" for biochemical compounds with antitumorigenic properties is 100 times higher for marine than for terrestrial biodiversity.[15] Several products based on MGR have already hit the markets, such as the antiviral drug Carragelose®, developed from red seaweeds; Yondelis® for the treatment of soft-tissue sarcoma, derived from the extracts of ascidian tunicate; Prialt®, an analgesic developed from isolates of the *Conus magnus* cone snail or the anti-cancer drug Halaven®, a synthetic analogue of isolates from a Japanese sea sponge.[16] Chitin from marine organisms can be used to treat bacterial, viral and fungal infections; silica from sponges are employed as coatings for surgical instruments and in microelectronics; compounds of microalgae can encapsulate drugs for better delivery and several isolates of marine organisms promise leads for the treatment of HIV.[17] MGR hold special promise for industrial biotechnology. Microbes from hydrothermal vents in the deep sea are biologically highly diverse. They are adapted to extreme temperatures, pressure and environmental toxins, and their enzymes are applicable in areas ranging from detergents over wastewater treatment up to pulp and paper processing.[18]

ABS from MGR face several tricky legal challenges. MGR in territorial waters and Exclusive Economic Zones fall within the ambit of the CBD. For international waters, UNCLOS provides two separate regimes. Its part XI applies to the Area: the seabed, subsoil and ocean floor beyond territorial waters and Exclusive Economic Zones. This part of UNCLOS was arguably the thorniest issue during the drafting of the convention. It is the reason the US never ratified the convention.[19] Several other industrialized countries only decided to ratify after the 1994 implementing agreement on part XI had incorporated several amendments.[20]

Pursuant to article 136, "the Area and its resources are the common heritage of mankind," and no property or sovereignty claims over them are permissible.[21] Activities in the Area, including resource extraction, are to be "carried out for the benefit of mankind as a whole."[22] This raises the question of whether MGR in the Area are part of the common heritage. There is no easy answer to that. Technically, article 133(a) defines "resources" as "all solid, liquid or gaseous mineral resources *in situ* in the Area, including polymetallic nodules." Yet there is a protracted legal debate on whether the definition of "resources" as "mineral resources" is exhaustive and thus expressly excludes "non-mineral resources" such as MGR.[23] Moreover, while UNCLOS defines "resources" as minerals, the common heritage encompasses not only the *resources* of the Area but also the Area *itself*: insofar as MGR in the seabed, ocean floor or its subsoil constitute parts of the Area, they are part of the common heritage.

The second regime applies to the water column above the Area. Here, the Freedom of the High Seas entails the right of states (and vessels flying their flags) to freely engage in scientific research. Unlike the Area, the water column above it is *res nullius*: it is not the common heritage of mankind but simply belongs to nobody. Other states neither have the right to regulate or otherwise limit access to MGR in those international waters nor to demand the sharing of benefits. The Freedom of the High Seas has several important qualifiers. The freedom of scientific research is, *inter alia*, subject to UNCLOS part XIII, which obliges parties to promote international scientific cooperation and share resulting knowledge and data.[24] Article 241 furthermore holds that "[m]arine scientific research activities shall not constitute the legal basis for any claim to any part of the marine environment or its resources." In principle, this provision could be construed as prohibiting intellectual property claims on "parts" and "resources," including MGR, of the high seas.[25]

MGR in international waters are unambiguously within the scope of the Freedom of the High Seas. The more difficult question is whether the Freedom extends to MGR in the Area, or whether those fall under the Part XI common heritage regime.[26] Potentially two regimes thus apply to MGR in ABNJ, with MGR in territorial waters falling under state sovereignty and those in Exclusive Economic Zones under the sovereign rights of states to the water column and the continental shelf. Yet MGR do not neatly abide by the legal boundaries established by UNCLOS and the CBD. Fish stocks, for instance, may move horizontally among territorial waters, Exclusive Economic Zones and international waters. Other marine organisms may move vertically from the Area to international waters. This immensely complicates the practical application of the different regimes.

Situation structure

MGR are different from the other types of genetic resources discussed in this book: originating from ABNJ, there are no "providers," merely users. Again using patent claims by geographic origin as an indicator of the relative capacities to utilize those genetic resources and to create benefits, the emerging picture is strongly asymmetrical. The US, Germany and Japan alone account for 70% of all relevant patent claims worldwide. The Top 10 jointly account for 90% (figure 8.1). The largest share of those patents pertain to chemical and pharmaceutical inventions.[27]

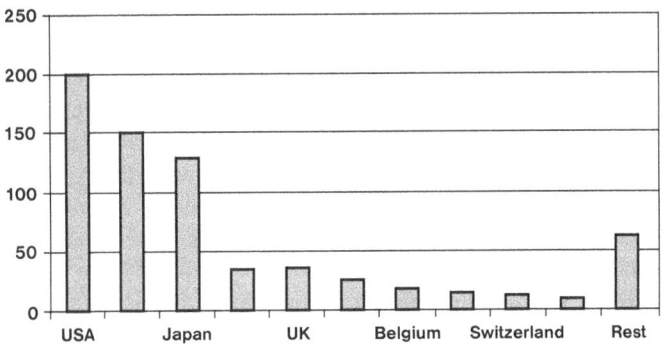

Figure 8.1 Patent claims on marine genes by origin
Source: Arnaud-Haond et al.

Despite those asymmetries, the absolute number of patents is still low and the total commercial value of MGR marginal, compared to the rest of the global bio-tech industry. For instance, despite some important pharmaceutical applications for MGR, no proper "blockbuster" drugs have yet emerged. This is not to say that MGR are commercially insignificant. The annual sales volume of just six drugs derived from MGR (Zovirax®, Combivir®, Trizivir® and the above-mentioned Yondelis®, Prialt® and Halaven®) amounted to US $648 million in 2015.[28] Using the FAO's SMTA as a benchmark, the annual benefits that would be shared on a multilateral basis just from those six products would amount to more than US $32 million per year.[29] While those numbers have significant uncertainties attached, a perfectly implemented ABS regime for MGR modeled on the FAO Seed Treaty could potentially generate a three-digit million dollar amount per year, which, in turn, could be mobilized for the conservation of marine and coastal ecosystems.

The public sector also plays a large role in the utilization of MGR. This includes applied science, which can lead to patent claims as well as basic science. Public re-search reflects itself in growing numbers of pertinent scientific publications, which rose from just over 200 in 2001 to close to 700 in 2010. The EU and the US jointly account for 60% of published papers and 68% of scientific citations. The next three in line, China, Japan and South Korea, each account for between 4% and 1% of publications and citations.[30] Beyond the public sector, numerous smaller enter-prises are active in blue biotechnology and MGR, including up-stream companies screening for active biochemical compounds to sell to larger manufacturers. From the side of the global corporate players, most major pharmaceuticals companies have division for marine biology where MGR are being used as well.[31]

Unlike for genetic resources stored in *ex situ* collections or collected by National Influenza Centers, access to MGR cannot be regulated effectively. The major barriers to sample acquisition are technology and expertise. MGR in coastal waters are relatively easy to access. Bioprospecting on the high seas is already more challenging, and sourcing deep-sea samples, for instance from the incredibly diverse microbial communities found in hydrothermal vents, is all but impossible except for a handful of actors, requiring specially designed submersibles or drones capable of operating at depths between 1 and 10 km, technology that is not widely available and is concentrated in the hands of few research centers. Operating a scientific vessel with a remote-controlled vehicle for sample extraction may cost up to US $5 billion per month.[32] The extraction of deep-sea MGR and their subsequent storage poses additional technological hurdles. These types of expeditions are predominantly in the hands of public research centers with the private sector contributing at best indirectly by contributing funding. Access by commercial actors thus takes place only at the stage where collected samples enter the laboratory.[33]

Being situated in ABNJ, there are no "providers" of MGR but rather users and non-users. Presently, there is neither a legal framework to regulate access to MGR nor even a remotely practical way to do so in principle. This makes the case different from PGRFA or viral materials, where the cooperation of provider countries is, to varying extents, required in order for access and subsequent utilization to occur. Presently, the only access barrier is technological and financial in nature. A small handful of states are able to extract and utilize MGR, whereas all others have neither the capacities to do the same nor the option of using access regulation to gain leverage in the negotiation of a benefit-sharing regime. From the four cases discussed in this book, here, interdependence is at its most asymmetrical. Yet this interdependence exclusively revolves around the distribution of benefits from marine bioprospecting and blue biotechnology, as there are no negative externalities arising from relevant intellectual property claims. As MGR in ABNJ under the control of individual states, moreover, any benefit-sharing regime will by necessity need to be multilateral rather than bilateral.

Negotiations

Here, I cover the deliberations within the Open-ended Informal Consultative Process on Oceans and the Law of the Sea (ICP), the Working Group on Marine Biodiversity Beyond Areas of National Jurisdiction (WG-BBNJ) and the later Preparatory Committee (PrepCom), as well as relevant processes under the CBD. Attempts to better protect marine biodiversity were initially unconnected from the question of ABS from MGR. The CBD had moved into the field early on with the 1995 Jakarta Mandate on the conservation and sustainable use of marine and coastal biological diversity through ecosystem-based management.[34] In 1999, the UN General Assembly established the ICP for facilitating the Assembly's annual review of developments in ocean affairs.[35] The ICP was to be an informal and deliberative setting without a negotiating mandate. In its three sessions from 2000 to 2003, the group debated fisheries, marine pollution, marine technology, anti-piracy measures, protection and preservation and integrated ocean management. The focus of the consultations shifted in 2002 after the World Summit on Sustainable Development called for measures on ocean management, including improved inter-agency coordination within the UN system, the better implementation of existing plans, targets and international treaties, improvements to the scientific understanding of marine and coastal ecosystems, and the creation of "representative networks" of Marine Protected Areas until 2012.[36] Later in 2002, the UN General Assembly extended the ICP's mandate and recommended that consultations focus on, *inter alia*, the protection of vulnerable marine ecosystems.[37]

The debate on MGR first commenced in the CBD yet quickly spilt over into the ICP. The 2003 meeting of the CBD's Subsidiary Body on Scientific, Technical and Technological Advice recommended to "compile and synthesize information on the status and trends of deep sea bed genetic resources,"[38] which was picked up by the subsequent COP in 2004, thus "focusing attention on the issue."[39] ABS from MGR as well as marine biodiversity would continue to be dealt with under both the CBD and the General Assembly in the following years. Over time, the ABS was absorbed exclusively by the latter, and the role of the CBD in marine biodiversity became limited to providing technical and scientific input.

Common heritage versus Freedom of the High Seas

The debate in the ICP quickly heated up when, at its fourth session in 2003, Mexico was the first to propose negotiations on rules for the commercial utilization of MGR in ABNJ. This opened the Pandora's Box of common heritage versus Freedom of the High Seas. The G77/China quickly came to embrace the common heritage approach to MGR in the Area,[40] with some members preferring to broaden the scope of ISA to biological resources.[41] Industrialized countries steadfastly refused to consider the application of common heritage to non-mineral resources. In 2004, the UN General assembly established the WG-BBNJ to "study issues relating to the conservation and sustainable use of marine biological diversity" in ABNJ.[42] The group's mandate included a review of existing governance arrangements and to indicate options for improving international cooperation. Meeting for the first time in 2006, the common-heritage debate quickly became an indispensable feature.[43] In the view of the G77/China, implementing common heritage would require developing operational rules on ABS, possibly under the ISA. The EU, Japan and the US all opposed the applicability of UNCLOS part XI to MGR, although the former admitted that voluntary guidelines for marine research might be put in place, and that some attempt should be made to clarify the legal status of MGR in the Area. With ABS from MGR being the "number one priority for the developing world,"[44] the G77/China was reluctant to move forward on Marine Protected Areas. The stance of developing countries was re-affirmed in a decision by CBD COP 7, meeting a month later and noting that "deep seabed ecosystems [...] contain genetic resources of great interest for their biodiversity value and for scientific research as well as for present and future sustainable development and commercial applications."[45] Parallel to the WG-BBNJ, the issue continued to haunt the ICP process as well. At ICP 8 in June 2007, some developing countries pushed for the drafting of legally binding rules.[46] Yet the disagreement over the legal status of MGR led to failure in producing a negotiated text on possible elements of a new biodiversity regime for consideration by the UN General Assembly.[47]

Governance and implementation gaps

ABS from MGR turned out to be only one of several controversial items being addressed in the WG-BBNJ. From the beginning of the Working Group, a key dispute was the question of governance and implementation gaps. The notion of a "governance gap" referred to the alleged insufficiency of existing institutional arrangements resulting from "the absence of institutions or mechanisms at the global, regional and sub-regional levels and from inadequate mandates of existing organizations and mechanisms."[48] "Implementation gap," conversely, denoted the insufficient implementation of existing treaties, plans and targets.[49] Suggestions included improving flag state control over their vessels in ABNJ and better port state control or performance reviews of Regional Fisheries Management Organizations.[50]

Some actors saw both an implementation *and* a governance gap, others saw either, and still others saw none. The EU quickly emerged as the most-vocal proponent of a new implementing agreement for UNCLOS for closing the governance gap and for establishing Marine Protected Areas in ABNJ. Such Areas traditionally referred to sectoral measures yet, over time, have become largely synonymous with the ecosystem-based approach, which, while defined in different ways, broadly refers to the comprehensive, integrated management of ecosystems on the basis of science and considerations of equity and socio-economic aspects.[51] For the EU and others, an implementing agreement would contribute to the protection of marine biodiversity in ABNJ by operationalizing some of UNCLOS' general and unspecific provisions, by enhancing the coordination between the various multilateral and regional institutions that are, to different extents, competent for biodiversity in ABNJ and provide for Environmental Impact Assessments for activities that might place it under threat.[52]

Australia, Iceland, Japan, Mexico, Norway and the US did not consider an implementing agreement necessary and proposed focusing on the implementation gap instead.[53] For Norway and the US, there was also skepticism on whether a new, international instrument on Marine Protected Areas would interfere with the Freedom of the High Seas: an area-based management tool for biodiversity in ABNJ could have far-reaching implications, for not only fisheries but also deep sea mining and shipping.[54]

The CBD: from competition to division of labor?

Broad support for Marine Protected Areas in ABNJ had been building in the CBD since 2004 when an Ad Hoc Open-ended Working Group was created to, *inter alia*, "explore options for cooperation for the establishment of marine protected areas in marine areas beyond the limits of national jurisdiction."[55] Two years later and mere weeks after the WG-BBNJ had convened for its first session and began haggling over whether or not a governance gap exists in ocean affairs, CBD COP 8 formally recognized both the necessity of a comprehensive, ecosystem-based approach to Marine Protected Areas in ABNJ as well as the CBD's "key role in supporting the work of the General Assembly" through the provision of relevant information and advice.[56] The Ad Hoc Open-ended Working Group on Protected Areas subsequently devised scientific criteria for identifying areas in special need of protection, which were formally adopted by the COP in 2008 and resulted in a preliminary list of Ecologically or Biologically Significant Marine Areas in 2012.[57] The Aichi Targets, adopted at the 2010 Nagoya Summit, established both the 10% protection target for Marine Protected Areas and the application of the ecosystem-based approach to aquatic plants, fish and invertebrate stocks until 2020.[58]

Many parties, particularly Canada, Iceland, Japan, New Zealand, Norway and South Korea, regarded the CBD discussions on Marine Protected Areas with skepticism, concerned that CBD overreach could undermine the parallel debates in the WG-BBNJ and initially opposing the designation of Marine Protected Areas.[59] Being primarily concerned about Illegal, Unreported and Unregulated fishery and destructive fishing practices, those large fishing nations preferred the process under the General Assembly due to the availability of issue linkages with their own priority issues. The G77/China were equally unenthusiastic, being primarily interested in ABS from MGR. Devising an MGR regime under the CBD faces formidable legal problems; in addition, the CBD's negotiations on ABS and on Marine Protected Areas proceeded on separate tracks, thus limiting the scope for linkages.[60] The main players pushing for a strong role for the CBD were the EU and Australia.[61]

The relationship between the two processes was gradually refined, with the WG-BBNJ becoming the focal point for discussions of marine biodiversity and MGR in ABNJ. In 2006, both COP 8 and WG-BBNJ 1 recognized "the United Nation's General Assembly's central role" in addressing biodiversity in ABNJ.[62] The annual General Assembly resolution on ocean affairs passed later that year "gave the final touches to the division of labor that had been progressively emerging among international organizations and processes related to marine biodiversity beyond national jurisdiction and high seas protected areas."[63]

Yet discussion in the CBD was not limited to conservation. As noted above, the CBD's Subsidiary Body on Scientific, Technical and Technological Advice had commenced discussions on MGR in ABNJ already in 2003. In 2006, two years after negotiations on what would become the Nagoya Protocol had officially been launched, several developing countries started pushing for ABS from MGR within the CBD.[64] Yet only in the final stretch of the negotiations did MGR become a contested topic. The African Group was particularly vocal in advocating that MGR be included within the scope of the instrument.[65] Under article 4(b), the CBD allows contracting parties to regulate bioprospecting activities carried out by their own nationals in ABNJ, although its article 15 provisions only apply to genetic resources falling under national sovereignty.[66] The Nagoya Protocol eventually limited itself to those genetic resources within the narrower scope of

CBD article 15 rather than the wider one under 4(b), thus ostensibly excluding MGR in ABNJ. As with many issues related to the treaty text, this exclusion has been debated. The Protocol contains an enabling clause for the creation of a global, multilateral benefit-sharing mechanism under article 10: parties "shall consider the need for and modalities of" such a mechanism for genetic resources in transboundary situations or where Prior Informed Consent cannot be granted or obtained. Article 10 includes, for instance, genetic resources with multiple countries of origin or "orphan" genetic resources for which the provider cannot be identified, for instance due to lacking documentation in *ex situ* collections.

Yet some parties have made the point that MGR in ABNJ fall into the scope of article 10 as well, since the absence of national jurisdiction makes it impossible to "grant" or "obtain" Prior Informed Consent for accessing them. In 2012, COP 11 launched a consultative process to identify the need and possible options for implementing the article.[67] In subsequent meetings, several developing countries have argued for including MGR within the discussions on article 10, with Canada and the EU cautioning against renegotiating the Protocol's geographic scope.[68] Yet in the view of the African Group, not only should MGR be included in the article 10 deliberations, but also their utilization should be subject to the Protocol's monitoring and compliance provisions.[69]

Ultimately, article 10 only creates an obligation on parties to deliberate the necessities and modalities of a mechanism for multilateral benefit-sharing, not to actually deliver it. Any ABS regime for MGR under article 10 would also restrict the ways in which shared benefits could be used, that is, for supporting the conservation and sustainable use of biological diversity. Moreover, the article does not specify whether such a mechanism should necessarily be created within the overall framework of the CBD. Thus, the WG-BBNJ process, as it developed from 2011 onward (see below), can be considered as satisfying the article 10 requirement. So why did some industrialized countries oppose the consideration of ABS from MGR in the context of article 10? At least in the case of the EU, one possible answer is that the debate was to be confined to the WG-BBNJ where it could be used as a bargaining chip for garnering the support of the G77/China for the EU's own policy objectives in regards to the protection of marine biodiversity. Understood in this way, the timing is not coincidental: the EU opposition to discussions on MGR under the CBD was to facilitate the process that was simultaneously unfolding under the auspices of the UN General Assembly.

Towards a new international agreement

Already in 2007, there were signs that the EU was willing to accommodate the G77/China in order to get support for an UNCLOS implementing agreement.[70] In May 2008, the EU proposed to the second meeting of the BBNJ that the group develop a multilateral ABS regime for MGR akin to the Seed Treaty, breaking sharply with its earlier insistence on the unequivocal applicability of the Freedom of the High Seas. This was part of the larger convergence of bargaining positions among the EU, the G77/China and a few others. Japan and the US continued to oppose an ABS regime and, seconded by Canada and Norway, a possible UNCLOS implementing agreement. Thus, besides MGR, the governance gap/implementation gap dispute persisted.

The breakthrough came at WG-BBNJ 4 in 2011 when parties agreed to a comprehensive package that would form the basis of further debates.[71] Based on a joint EU-G77/China initiative, the working group agreed to proceed simultaneously on ABS for MGR (including the possibility of monetary benefit-sharing), capacity building, the transfer of maritime technology, as well as Marine Protected Areas and Environmental Impact Assessments.[72] This linkage itself fed into the broader processes leading up to the 2012 Rio +20 summit intended to reinvigorate the political commitments to sustainable development made 20 years earlier at the Rio Earth summit. While oceans are the "quintessential sustainable development issue," the implementation of the targets and objectives agreed at the highest political levels both in Rio 1992 and at the 2002 Johannesburg World Summit on Sustainable Development had largely been sluggish.[73] At Rio +20, governments committed to "address, on an urgent basis, the issue of the conservation and sustainable use of marine biological diversity of areas beyond national jurisdiction including by taking a decision on the development of an international instrument under UNCLOS,"[74] thus reinforcing the overall political momentum.

A year later, the UN General Assembly requested that the BBNJ-WG "make recommendations [...] on the scope, parameters and feasibility of an international instrument under [UNCLOS]" and mandated three more meetings in 2014 and 2015.[75] Based on the WG-BBNJ's final recommendations, the General Assembly decided, in 2015, to establish a legally binding instrument on the conservation and sustainable use of marine biodiversity in ABNJ. To that end, a Preparatory Committee (PrepCom) would negotiate a draft text during 2016 and 2017 with the General Assembly subsequently deciding on whether or not to convene a diplomatic conference.[76] At the time of writing, the second session of the PrepCom just concluded. Despite the considerable political support and decisions made at the highest political levels, the process is fraught with various uncertainties and frictions. Many of those do not relate to ABS from MGR but to the broader issue of marine biodiversity. A major problem continues to be the role of Marine Protected Areas, particularly in regards to fisheries. As Regional Fisheries Management Organizations are already able to designate Marine Protected Areas in international waters, in a way that is binding on all contracting parties to the UN Fish Stocks Agreement, what would be the added value of a new, global agreement for designating such areas? And would it include some kind of enforcement component? For some countries, the agreement raises the risk of undermining the work of Regional Fisheries Management Organizations.[77] For others, excluding activities that are presently regulated through sectoral approaches would defeat "the very purpose of the exercise."[78] Moreover, should high-seas fisheries even be included in the agreement? Being the primary source of marine biodiversity loss, there is a strong case to be made that it should, yet countries like Iceland, Japan, Norway and Russia consider this a threat to their economic interests.

In regards to ABS, first, while some governments (such as China and Norway) appear to be moving away from the fundamental dispute over common heritage and Freedom of the High Seas and opting for pragmatic solutions instead, many others (such as the African Group) continue to frame the debate in precisely those terms.[79] The question of common heritage directly relates to the perennial dispute over intellectual property rights. In the cases discussed in Chapters 5 to 7, intellectual property claims over genetic resources were frequently understood as infringing on the principle of national sovereignty. In ABNJ, common heritage provides the normative grounds for attempting to link intellectual property claims to a benefit-sharing obligation towards the international community. Industrialized countries predictably consider the PrepCom to be the wrong body for discussing intellectual property.[80]

Second, there is the debate on the regime's scope. Some developing countries prefer to bring both the MGR of the Area as well as those of the high seas under the ABS regime and, potentially, the common heritage principle. While UNCLOS part XI can be construed as encompassing MGR within the seabed, the ocean floor and its subsoil, those in the water column above it fall unambiguously within the ambit of the Freedom of the High Seas. Recognizing them as the common heritage of mankind would thus require major amendments to UNCLOS. A different issue is whether fish constitute MGR—in all likelihood, attempts to create an obligation to share the benefits from the utilization of fish would blow the negotiation process right out of the water. And then there is the problem of the geographical delimitation with the CBD/Nagoya regime. The latter encompasses states territorial waters and Exclusive Economic Zones, whereas a new ABS regime for ABNJ would cover the area beyond. Inconsistencies can arise when marine organisms travel horizontally between the two zones:[81] in that case, the applicable legal regime may not be readily identifiable, or users might have incentives to misrepresent the precise geographical point at which materials were acquired in order for the "weaker" regime to apply.

Third, how to delineate the commercial utilization of MGR from marine scientific research, which possesses a distinct legal status under UNCLOS? At the point of sample collection (on the high seas or in the Area), the two are indistinguishable—in fact, bioprospecting activities are almost exclusively being carried out by public research institutions, with commercial utilization of MGR only commencing once samples have been collected.[82] As with all other ABS regimes, there is the danger of undermining non-commercial research if access to genetic resources becomes over-regulated in order to extract benefits from commercial utilization—a point China and Russia have made. For Japan, new regulations on marine scientific research would be "in principle unacceptable."[83] The question of the points in the value chain, from initial collection to the laboratory and the end users, where regulation is to take place, including the triggering of a benefit-sharing obligation, is far from trivial.[84]

Finally, then, there is participation and the agreement's legal nature. Despite the mandate by the UN General Assembly, some countries, such as Canada, South Korea, Russia and the US, continue to favor a non-binding approach.[85]

The US and several South American countries not being members of UNCLOS, moreover, an issue is whether the agreement should be concluded under the Convention or as a free-standing agreement, the latter option allowing for broader participation.

Many of those questions will probably be resolved until the end of 2018 when the General Assembly will decide on whether to convene a diplomatic conference. Until then, the contours of the agreement, as well as its ABS component, are only slowly taking shape. Yet looking at the process so far, we may nevertheless attempt to reach some tentative conclusions of how to conceptualize its broader linkages to the genetic resources regime complex.

Explaining the negotiation process

The lack of institutional outcomes limits the extent to which the interplay management, regime shifting and situation-structural hypotheses can be tested, yet the process so far gives some clues as to each approach's relative strengths and weaknesses. While conservation and sustainable use of marine biodiversity in ABNJ and the question of ABS from MGR form part of a package, I limit myself to the latter, considering that the issue area of marine biodiversity conservation is substantially different from the focus of this book.

Interplay management

An ABS regime for MGR can be conceived of as an attempt to resolve important inconsistencies between UNCLOS and the patent regime. Those inconsistencies pertain to MGR in the Area, marine scientific research and the benefit-sharing provisions of the CBD as well as relevant passages of UNCLOS itself.

First, if we consider MGR of the ocean floor, the seabed and its subsoil as "parts" of the Area, they are the common heritage of mankind and shall not be appropriated by states, natural or juridical persons.[86] Patent claims on MGR from the Area would thus not be permissible under UNCLOS and create conflicts with the relevant intellectual property rights regimes. UNCLOS' prohibition of appropriating parts of the Area is absolute—that is, it does not depend on whether or not benefit-sharing arrangements for the appropriated MGR would be in place. This is different for mineral resources, which may be appropriated (or "alienated") in line with UNCLOS part XI and the regulations of the International Seabed Authority. Legal conflicts from the appropriation of the common heritage of mankind through intellectual property rights can thus be avoided if the Authority were to be granted the right to regulate the conditions under which MGR from the Area are collected and utilized, as is already the case for mineral resources. This would reconcile common heritage with intellectual property, an option that figures prominently in the negotiation process.

Second, a potential inconsistency relates to marine scientific research, which "shall not constitute the legal basis for any claim to any part of the marine environment or its resources."[87] The problem here is reminiscent of the Seed Treaty's prohibition of patents on PGRFA "in the form received":[88] if patent protection is claimed for MGR that have not undergone sufficient biotechnological innovation, they may still be considered a "part" or a "resource" of the marine environment and, considering that such an invention would inevitably be based on "marine scientific research," not be claimable in line with UNCLOS.[89] This would lead to a similar conflict as in the case sketched out in the previous paragraph. Those provisions apply both to MGR in the Area as well as those of the high seas: even where bioprospecting takes place under the Freedom of the High Seas, this freedom is subject to the part XIII provisions on marine scientific research.[90]

Finally, the move towards an ABS regime for MGR could avoid tensions between private appropriation and the UNCLOS article 140 proscription that activities in the Area are to be "carried out for the benefit of mankind as a whole." In fact, it "would be a significant anomaly if bioprospecting and the sustainable use of biodiversity generally were to be the only activities in the Area not undertaken for the benefit of mankind as a whole."[91] Marine scientific research in the Area (but not in the water column above it), moreover, is subject to special provisions, as contracting parties shall promote cooperation by developing international research programs to specifically benefit developing countries and "effectively disseminate the results of research and analyses when available."[92] It is also noteworthy that the CBD's benefit-sharing objective applies, within certain limits, to MGR in ABNJ, both in the Area and the high seas. As set out in article 1 of the Convention, the objective is not limited to genetic resources under national jurisdiction. Rather, the CBD also applies to "processes and activities," arguably including bioprospecting, in international waters that fall under flag state jurisdiction.[93] Here, contracting parties may be under an obligation to ensure that benefits from the utilization of MGR are shared fairly and equitably. While this excludes both contracting parties' obligations under CBD article 15 and the entire Nagoya Protocol, the case can be made that there is a general obligation to share the benefits arising from the utilization of MGR. The push towards an ABS regime for MGR thus satisfies this requirement under the CBD and improves on the status quo, which only increases "disparities in the distribution of global welfare and economic wealth," as the G77/China notes.[94]

A major challenge negotiators will need to address, however, is whether the ABS regime will apply solely to the MGR of the Area or also to those of the high seas. A comprehensive regime covering both regions would generate tensions with the Freedom of the High Seas—rather than a mere implementing agreement, this would arguably require the partial amendment of UNCLOS part VII; yet an ABS regime that differentiates between the Area and the high seas is impractical due to the vertical movements of many marine organisms between the two zones. Thus, despite a comprehensive regime being legally (and politically) challenging, "from a scientific and managerial point of view, a comprehensive regime covering the entire ABNJ appears to be indispensable."[95]

It is noteworthy that a major actor, the EU, only recognized the need to assess and clarify the legal status of MGR once the possibility of a package deal came in reach. In fact, until the 2008 meeting of the WG-BBNJ, only developing countries were pushing for the formal acceptance of MGR as the common heritage of mankind. While the closing of the alleged governance gap was the chief motivating factor for the EU's engagement with the negotiation process, the potential governance gap in regards to MGR appears to have been a matter of political expediency and not principled concern over unresolved tensions between UNCLOS and the patent regime.

In the relationship between the emerging regime for MGR and the Nagoya Protocol, the latter establishes a relatively clear division of labor by exclusively covering genetic resources within the scope of article 15 CBD and thus excluding all materials found in ABNJ.[96] In the ongoing implementation phase, the African Group, for which the inclusion of genetic resources in ABNJ had long been of "highest priority,"[97] is attempting to use the article 10 obligation to consider a global multilateral benefit-sharing mechanism for extending the Protocol's geographical scope. While the CBD's benefit-sharing obligation does not presuppose the granting of Prior Informed Consent and the establishment of Mutually Agreed Terms, and while article 4(b) CBD arguably extends the obligation to ABNJ, the African Group's legal case for applying article 10 of the Protocol beyond national jurisdictions is weak.[98] As other actors generally advocated against the inclusion of MGR from ABNJ in the Protocol's scope (or were largely indifferent, at the least),[99] this is an important data point in favor of interplay management.

Regime shifting

A different way of looking at common heritage is as a counter-regime norm intended to weaken both the international patent regime and the Freedom of the High Seas regime, and/or to strengthen the coverage of the common heritage approach in Area and extend it to MGR in international waters. Considering the anti-patent agenda, which several developing countries have been pursuing in other forums, the explanation intuitively appears plausible. The "usual suspects," Brazil and India, are among those members of the G77/China who continue to hold on to the formal recognition of the MGR of the Area as the common heritage of mankind and who insist on discussing the role of intellectual property rights in this regard.[100]

As noted multiple times already, whether the MGR of the Area fall under common heritage is a tricky legal question to which there is no unambiguous answer. Yet many members of the G77/China do not limit themselves to demanding the application of common heritage to MGR in the Area, but rather seek its extension to the high seas as well. This, in turn, would roll back the Freedom of the High Seas and the concomitant right to appropriate MGR through intellectual property rights without restrictions beyond the part XIII provisions on marine scientific research.

The argument has some merit, yet is weakened by the apparent willingness of several members of the G77/China to find a compromise solution on ABS from MGR that would not require a decision on whether they fall under common heritage or the Freedom of the High Seas. Mexico has repeatedly proposed such a pragmatic approach since 2006.[101] While common heritage is still the official position of the G77/China,[102] several of its members, including China and Indonesia, have proffered the pragmatic approach as of late.[103] The timing suggests that, as negotiations on the international instrument are picking up pace, some developing countries might be weakening their stance on the issue in order not to jeopardize the broader bargaining process and might be willing to trade common heritage in exchange for tangible gains under a pragmatic benefit-sharing agreement.

Situation structure

Looking only at MGR and blue biotechnology while disregarding the linkage to the broader issue of conservation and sustainable use of marine biodiversity in ABNJ, user countries should strictly prefer the status quo. Unlike the cases of the Seed Treaty or the WHO Framework, user countries require no international cooperation whatsoever in order to obtain access to the MGR of the high seas or the Area. In fact, any cooperative arrangements that regulate access or impose benefit-sharing obligations are costly to the nascent blue biotech industry. While this sufficiently explains the opposition to an ABS regime by the US and Japan, the first and second largest users of MGR in the world, it cannot account for the EU's willingness to compromise on the issue. In fact, the European Commission's Blue Growth strategy considers blue biotechnology a focal area with high potential for technological innovation and economic growth, which could develop into a mass market by the late 2020s.[104] The EU thus appears to be willing to channel large amounts of resources through an ABS regime for MGR in the future.

There are two ways of explaining this behavior. The first, which also bears relevance for the Seed Treaty, has to do with the problem that this book has treated the EU as a unitary actor. Yet EU foreign policy is frequently an aggregation of quite diverse interests of both its member states and supranational institutions. The only member states with substantial innovative capacities in MGR and blue biotechnology are Germany, the UK, France and, lesser so, Denmark, Belgium and the Netherlands (see section 8.2 above), whereas others presently have nothing to lose from an ABS mechanism. I return to this issue in Chapter 9.

More in line with the situation-structural hypothesis is that negotiations on the ABS regime are linked to the protection of marine biodiversity in ABNJ, the core of the EU's bargaining position since 2007. Considering that most other industrialized countries have been, and still are, skeptical about the merits of a new international instrument for marine biodiversity, the linkage to ABS enabled the formation of a broad EU-G77/China coalition. To put it differently: the support that the G77/China are giving to the EU on the issue biodiversity conservation and sustainable use holds the prospect of developing a new ABS mechanism on the foundations of UNCLOS. Whether or not such a mechanism formally incorporates the common heritage approach is peripheral as long as tangible gains can be expected from the eventual regime.

The club scenario

If some parties decide that the process under the General Assembly is unlikely to meet their respective interests or requires too many concessions or linkages to bring to a conclusion, what are the prospects for club cooperation instead? As the collection and utilization of MGR is out of the control of non-users, and as access to international waters cannot be regulated effectively by third parties, a non-user club would make remarkably little sense. Similarly, users have little reason to enter into club cooperation themselves: the status quo, in which access is unregulated, is preferable over any outcome in which benefits would have to be shared and benefit-sharing possibly even supported by compliance measures. A club would also fail to realize the precise reason users are, in principle, open to the idea of having ABS from MGR in the first place: using it as leverage for achieving other objectives related to the conservation and sustainable use of marine biodiversity.

Notes

1 Druel (2011), 5.
2 Barkin and DeSombre (2013).
3 Druel (2011); Rochette et al. (2014), 33–34.
4 Crowder and Norse (2008), 772.
5 Besides the Fish Stocks Agreement, UNCLOS part XI is the second currently existing implementing agreement, see Harrison (2011), 85–114.
6 CBD article 4(b).
7 Kim (2013), 14–19.
8 Hart (2008), 3–4.
9 Kim (2013), 2.
10 Crowder and Norse (2008).
11 A/61/65, para 54.
12 Drankier (2012).
13 Kim (2013), 12–13.
14 CBD articles 8(a) and 4(b); see also Drankier (2012), 296–299.
15 de la Calle (2009), 210.
16 de la Calle (2009), 213.
17 OECD (2013), 30–34; see also Leary et al. (2009).
18 Leary (2007), 158–164.
19 Harrison (2011), 47–48.
20 de Marffy-Mantuano (1995); Tanaka (2012), 178–184.
21 UNCLOS article 137.
22 UNCLOS article 140.
23 Drankier et al. (2012), 402–403; Broggiato et al. (2014), 4–5.
24 UNCLOS articles 242 and 244.
25 Drankier et al. (2012), 396–397.
26 Hodgson et al. (2014), 48.
27 Leary et al. (2009), 190.
28 data from http://www.evaluategroup.com/, accessed 25 August 2016.
29 Under the SMTA's article 6.11 payments scheme.
30 ECORYS et al. (2012), 11–12.
31 ECORYS et al. (2012).
32 Broggiato et al. (2014), 2–3.

33 Leary (2007), 165–169.
34 Decision II/10.
35 A/RES/54/33.
36 Johannesburg Plan of Implementation, paras 32 to 36.
37 A/RES/57/141, para 62.
38 Recommendation VIII/3.
39 ENB (2004), 8.
40 ENB (2004), 3–4.
41 A/61/5, para 29.
42 Resolution 59/24, para 73.
43 ENB (2006b), 3.
44 ENB (2006b), 8.
45 Decision VIII/21, para 1.
46 ENB (2007), 3.
47 ENB (2007), 6–7.
48 A/63/79, Annex, para 43. See also Druel and Gerde (2014).
49 ENB (2006b), 3.
50 A/63/79, Annex, para 40.
51 Kim (2013).
52 Hart (2008).
53 ENB (2006b), 4.
54 Blasiak and Yagi (2016), 213.
55 Decision VII/28, para 29(a).
56 Decision VIII/24, paras 39 and 42.
57 Decision XI/17, Annex.
58 Aichi Targets 11 and 6; see also Rochette et al. (2014).
59 ENB (2005), 3–4 and 9; Morgera (2007), 7; Buck (2008), 252–254.
60 Morgera (2007), 6.
61 ENB (2005), 3–4.
62 Decision VIII/24, 7; A/61/65, Annex I, para 2.
63 Morgera (2007), 8.
64 ENB (2006c), 14–15.
65 ENB (2012), 10; Wallbott et al. (2014), 44.
66 CBD article 4(b); Broggiato (2011), 38.
67 Decision XI/1, part B.
68 ENB (2014b).
69 African Group (2012).
70 ENB (2007), 11.
71 Blasiak and Yagi (2016), 212.
72 A/66/119, Annex, para I(b).
73 Cicin-Sain et al. (2011).
74 Para 162.
75 A/RES/68/70, para 198.
76 Resolution 69/292.
77 Japan (2016).
78 New Zealand (2016).
79 ENB (2012); ENB (2014a), 1 and 4.
80 i.e., Switzerland (2016).
81 Broggiato et al. (2014).
82 Leary et al. 2009.
83 Japan (2016).
84 Tvedt (2013).
85 Rochette et al. (2015).
86 UNCLOS articles 136 and 137(1).

87 UNCLOS article 241.
88 Seed Treaty, article 12.3(d).
89 MGR may constitute such resources for the purpose of marine scientific research, which is addressed under UNCLOS part XIII; the definition of 'resources' as *mineral* resources only applies to part XI; see article 133.
90 UNCLOS article 87.1(f).
91 De la Fayette (2009), 269.
92 UNCLOS article 143.
93 CBD article 4(b).
94 G77/China (2016), sic.
95 Broggiato et al. (2014), 4.
96 Oberthür and Pozarowska (2013).
97 Wallbott (2014), 119.
98 African Group (2014).
99 Wallbott et al. (2014).
100 Brazil (2016); India (2016).
101 i.e., ENB (2006b), 8.
102 G77/China (2016).
103 ENB (2014a), 4; ENB (2015), 10; ENB (2016), 8.
104 COM (2012) 494 final, 12.

9 Conclusions

The cases discussed in this book show that situation structure can adequately explain most, yet not all, institutional changes in the genetic resources regime complex. As such, it offers a broader theoretical perspective than either interplay management or regime shifting. Both of the latter can partially explain different aspects of institutional change in ABS—yet the question is why, at times, institutional outcomes can be explained in terms of the former and at other times in terms of the latter. Why would governments engage in interplay management, say, in regards to the relationship of the Seed Treaty, the CBD and the CGIAR, yet opt for inconsistency in regards to the relationship between the Treaty's article 12.3(d) and the patent regime? If article 12.3(d) is a counter-regime norm directed towards patents, why does article 13.2(d)(iv) accept patent claims under the condition that benefits are shared? If Farmers' Rights were intended to balance the illegitimate appropriation of PGRFA by a handful of multinational seed companies and research institutes, why did the participants to the International Undertaking adopt the Farmers' Rights amendment at the same time as they conceded the Undertaking's compatibility with plant-variety rights? And why did they surrender large parts of their bargaining position on Farmers' Rights before the negotiations on the Seed Treaty were even picking up pace?

What the one approach cannot explain, the other can—but this raises the issue under which conditions each of them actually obtains! While not free from problems of its own, situation structure offers the prospect of a unified framework for explaining the link between regime complexity and institutional change. I conclude this book with several broader considerations on both the situation-structural approach to institutional change and ABS. I first turn to the major challenges the approach faces empirically before turning to the question of generalization. I conclude with several remarks on the future of ABS governance.

Limitations of the approach

Situation structure fares well and, in many instances, better than its competitors in explaining institutional changes in the genetic resources regime complex. As no single theory can explain everything, this is initially a satisfying result. Situation structure holds the promise of large inferential leverage—as a rather minimalist theoretical approach, it is potentially applicable to a broad range of other cases, a point to which I return below. Before doing so, I will look into its shortcomings and the lessons that can potentially be gleamed therefrom.

The biggest problem the approach faces is that, in some cases, it cannot account for the empirically observable variations in institutional preferences between actors that hold similar positions in the situation structure. More precisely, this concerns the difference in bargaining behavior between the EU and Switzerland, on one hand, and biotechnological powerhouses such as Japan and the US, on the other. Those actors possess the strongest biotechnological industries in the world in the seed sector, pharmaceuticals and, with the exception of Switzerland, blue biotechnology. Yet across the four cases, their willingness to seek compromises with developing countries, specifically the G77/China, differs. The US is party to neither the CBD nor the Seed Treaty and has consistently disputed the need for an international anti-biopiracy regime, an ABS regime for MGR and the UNCLOS implementing agreement within which the latter would be embedded. Japan only joined the Seed Treaty 12 years after its conclusion, has been one of the most outspoken critics of international measures against biopiracy and is also highly skeptical of the UNCLOS process. The EU and Switzerland, conversely, have offered various proposals for an international anti-biopiracy regime and have both played proactive roles in the negotiations on the Nagoya Protocol. The EU is the driving force behind a comprehensive UNCLOS implementing agreement that would include an ABS regime for MGR and has also met developing countries halfway on the question of Farmers' Rights in the negotiations on the Seed Treaty.

In the case studies, I have discussed several ways of explaining those differences. This includes the pre-existing disclosure regimes in several EU member states and Switzerland, which would have limited the legislative and regulatory changes necessary for complying with an international obligation to require disclosure of origin. It also includes differences in intellectual property law: As Japan and the US are virtually the only countries in the world that allow for patents on plant varieties, their opposition to the Seed Treaty is not surprising. While the scope of patentable subject matter has been gradually expanded for contracting parties to the European Patent Convention over the last decades, it is still comparatively narrower than for Japan and the US. In the case of the Pandemic Influenza Preparedness Framework, finally, there was no distinctively "European" way of bridge-building towards the G77/China. Instead, the revealed preferences of all industrialized countries were practically identical.

The second problem is more practical in nature. The case studies have spoken of "the" EU as if it were a unitary actor—which it is not. In terms of situation structure, significant differences exist among the various EU member states. Germany, Spain, the Czech Republic or the UK, at the time of writing still a member state, possess larger biotechnological capacities than, say, Poland, Portugal or Luxembourg. Yet most of the negotiations discussed here fall into the partial competence of the Union and the rest, such as those in the TRIPS Council, into its exclusive competence. The EU's bargaining behavior is thus the result of complex, internal coordination processes among member states and the supranational institutions. Even where individual member states retain formal competences, there are strong social, political and legal pressures on them to adopt common bargaining positions. For instance, where only few EU member states are bound to incur losses from a benefit-sharing regime for MGR (the majority is simply indifferent to the matter), the willingness of the EU and its member states to enter into compromises with the G77/China appears less surprising than if we assume "the EU" to be a homogenous actor with one particular position in the situation structure. A theory that intentionally neglects the domestic sources of foreign policy in order to achieve greater parsimony, such as the theory developed and tested in this book, will inevitably fall short in this regard.

The question is then: how far can theoretical simplification, for the purpose of increasing inferential leverage, tolerate unaccounted variation in the object to be explained? Or, to put it another way: if the usefulness of a theory is measured by the ratio between theoretical assumptions and the phenomena it can explain, how large can the unexplainable phenomena get before the theory ceases to be useful?

Generalizing the approach

As noted in Chapter 2, situation structure promises to explain both club cooperation and institutional layering as distinct types of institutional change in regime complexes, yet a comprehensive test of the approach requires cross-case variation on the dependent variable. Here, I will limit myself to some rather *ad hoc* remarks on how situation structure could be of use in explaining changes in *other* regime complexes. Those considerations are brief and tentative. At best, they should be seen as a preliminary outlook.

To recap, we should expect club cooperation to obtain instead of institutional layering if reformist actors can realize gains from cooperation in excess of any costs resulting from the non-participation of status quo-oriented actors. This means that the former are able to contribute sufficient goods for cooperation to be pareto superior, or that the latter do not generate unavoidable negative externalities. Let us look at the case of hazardous waste once more. From the 1980s onward, numerous developing countries began calling for a global ban on North-South waste exports.[1] The 1989 Basel Convention provides a global regime regulating the generation of and trade in hazardous waste yet aims to oversee and control trade, rather than banning it.[2] The Convention also suffered from a well-known loophole, by which waste exports intended for recycling instead of disposal do not fall under its control obligations—which, in the years after its adoption, led to waste exports declared for recycling increasing roughly in the same amount as waste exports declared as for disposal decreased, with exporters effectively changing the labels on the shipping containers and not much else. Efforts to tighten the Basel regime have failed for many years due to opposition from large exporters in combination with procedural disputes. During the 1990s, numerous regional treaties were put in place in Central America, Africa, the South Pacific and other areas, leading to the emergence of a regime complex for hazardous waste.[3] From the situation-structural perspective, what enabled club cooperation in the first place is the option of banning waste trade at the point of import. While export controls would enhance the implementation of a comprehensive ban, import controls by customs authorities.[4] Negative externalities from non-participants who choose to merely regulate, instead of ban, North-south waste shipments, can partially be avoided by club members, thus allowing reformist actors to bypass multilateral cooperation.

Situation structure also promises to account for within-case variation. Constraints from situation structure may pertain to some elements of a regime yet not to others. Climate change is such an issue area. On one hand, greenhouse gas emissions are unavoidable negative externalities. The countries with the largest emissions bear the highest costs of emissions reductions and, with the possible exceptions of India or Indonesia, are better equipped to adapt to the consequences of climate change than others. Those countries pushing most strongly for global emissions reductions, with a 1.5°C warming target until the end of the 21st century, are those that will face the highest costs: small island states that, under business as usual, are quite likely to disappear as a consequence of rising sea levels. In terms of emissions targets, those are the actors preferring the largest change from the status quo. The idea that small island states, as reformist actors, would enter into club cooperation amongst themselves for purposes of emissions reductions is obviously ludicrous yet illustrates the constraints situation structure may impose. The entire debate on "climate clubs," where some countries would bypass the UN process in order to cooperate more deeply on mitigation through exclusive settings, only deals with major emitters in a fairly commonsensical way. Yet, once we look at the theoretical literature of how actors that are dissatisfied with the status quo in international institutions behave, the slightly silly notion of Pacific island states teaming up to reduce their emissions in a way that is more ambitious than their (limited to non-existent) commitments under the Kyoto Protocol can elucidate the larger theoretical issue at stake. Yet beyond binding emissions targets, the climate regime has fragmented in an extraordinary manner in recent years. International initiatives on adaptation, emissions trading, technology transfer and so forth have proliferated immensely.[5] From the perspective adopted here, club members operating in those areas can realize gains regardless of whether others participate. Regional adaptation to climate change does not suffer from negative externalities if other countries do not join the club—the precondition being that club members have sufficient financial and technical resources to implement adequate adaptation policies.

Or take another instance of within-case variation: patents have been treated as a single regime in this book yet depending on one's perspective can be considered to constitute a regime complex of themselves. In the early 1990s, patents fell within the scope of WTO-TRIPS and a handful of WIPO treaties. Today, they figure on the agendas of several (other) UN organizations, including the CBD, the FAO and the WHO, where the relationship between patents and public health goes significantly beyond the narrower question of influenza preparedness and response discussed in Chapter 7. At the same time, most Free Trade Agreements concluded in the last two decades contain provisions on patents, usually requiring contracting parties to ratchet up their domestic patent law beyond the minimum standards proscribed in TRIPS—which is why those types of treaties are sometimes called "TRIPS-plus" agreements.[6] The politics of patents have been subject to much analysis.[7] It is striking that developing countries (as net importers of intellectual property) have chosen to operate through multilateral organizations for advancing their patent agendas, whereas industrialized ones (as net exporters) have tended to choose the route of club cooperation more often. A possible explanation is this: exporters of intellectual property can realize gains from cooperation regardless of what others are doing—by ratcheting up standards in those countries that become parties to Free Trade Agreements, the protection of their property is enhanced in the respective jurisdictions. The non-participation of others is not an issue here. Conversely, net importers cannot go below the minimum standards proscribed in TRIPS yet need the consent of exporters for softening-up the multilateral rules in areas such as agriculture or public health.

Finally, situation structure can possibly explain why fragmented governance architectures "with no identifiable core and weak or nonexistent linkages between regime elements"[8] are sometimes transformed into regime complexes, with one central institution taking on a coordinating role for the regulatory periphery. Consider the international regulation of Persistent Organic Pollutants, a class of chemicals that is transported around the entire globe through airwaves and water and accumulation in the food chain. Regulation has historically been sectoralized, with various regimes addressing, for instance, emissions or the ocean-dumping of Persistent Organic Pollutants and the broader chemical category of organochlorines. Yet the discovery, in the late 1980s, of the *extent* to which Persistent Organic Pollutants diffuse around the globe set in motion a political process for their comprehensive regulation.[9] This culminated, in 2001, in the conclusion of the Stockholm Convention, a multilateral treaty regulating the substances' production, trade, emissions and disposal. With Persistent Organic Pollutants emanating chiefly from Southern countries yet accumulating in the colder climates of the North, industrialized countries were chiefly pushing for the global instrument: as negative externalities from developing countries were effectively unavoidable, club cooperation would not have sufficed. Yet including developing countries within a comprehensive multilateral framework required both linkages and concessions: technical and financial assistance as well as the inclusion of several flexibilities, allowing contracting parties the continued usage of certain Persistent Organic Pollutants under specific conditions.[10]

All of those examples would deserve further investigation as to the type and the drivers of change in multi-institutional settings. Ultimately, situation structure could lead to a simple-yet-powerful theory that should allow for limited predictions of the ways in which institutional architectures in different areas will develop. Hopefully, this book is a useful first step down that road.

Looking ahead: the future of ABS governance

As this book dealt as much with regime complexity and institutional change as it did with genetic resources, I will conclude with some thoughts about the possible futures of ABS governance in light of the theoretical approach adopted here. Several important developments are presently underway that might have a decisive impact on the present governance structure as well as the underlying situation structure. For starters, industrial biotechnology is possibly gaining in significance in relation to the older, more-established fields of pharmaceutical and agricultural biotechnology. This implies that attention to microbial genetic resources, both terrestrial and from the oceans, could grow in the near future. The emerging ABS regime for MGR is a first hint that policy-makers could be turning their attention towards microbes and away from plant genetic resources—which are, by now, covered in a largely comprehensive manner by international treaties. Of all types of genetic resources, microbial ones presently have the largest economic potential, simply due to the fact that, unlike plant and animal biodiversity, very little is known about them. The distributive conflicts and situation structures in those areas are slightly different from those for other types of genetic resources. As discussed in Chapter 8, access to MGR cannot be regulated effectively, and very few actors have the capacities to access them in the first place. For terrestrial microbes, their arguably "cosmopolitan" (that is, even) spatial distribution[11] means that there is no clear separation between users and providers—"non-users" thus have little effective leverage for prodding users into fair and equitable benefit-sharing.

Another significant development is synthetic biology, which, in certain ways, could undercut the present governance architecture for ABS. At the least, it will require some adaptations to ensure that existing legal frameworks will be implemented effectively. Under the Pandemic Influenza Preparedness Framework, genetic sequence data is presently being treated differently than physical specimens—the utilization of the latter is subject to either SMTA 1 or 2, whereas utilization of the former is not; transfers of sequence data even fall outside the scope of the Framework's Traceability Mechanism.[12] Yet once technological changes increasingly allow for the rapid creation of synthetic viruses on the basis of digital blueprints, one of the two purposes of the Framework, the sharing of benefits, will quickly be undermined. The same goes for compliance: how, exactly, will contracting parties to the Nagoya Protocol implement their article 15 to 18 obligations on user measures in regards to genetic sequence data? Or, as an even more fundamental question, how far does sequence data even *constitute* genetic resources?[13] Those questions urgently need clarification as technological development progresses. There are also practical challenges: monitoring the transboundary movement of physical materials is difficult enough. Yet how do we devise an effective monitoring scheme for transmissions of digital data?

Another question is which new ABS regimes will emerge in the future and how they will shape the larger regime complex. At the time of writing, it appears likely that an ABS regime for MGR will be concluded as part of a wider UNCLOS implementing agreement. This development is particularly interesting because, as it appears, the new regime will formally recognize the genetic resources of the deep seas and, possibly, the high seas, as the common heritage of mankind. This would be a welcome development from a larger, normative perspective, as it would revitalize the governance of genetic resources under the perspective of humanity writ large instead of under the parochial interests of nation states and commercial users. It could also serve an important role in the area of microbial genetic resources more broadly, where the free and largely informal exchange of materials between non-commercial users might become threatened by access regulation.[14] The genetic resources of the Antarctic had briefly moved into focus during the negotiations on the Nagoya Protocol.[15] The biological diversity of the Antarctic is significant in terms of marine species and microbes yet is under threat from global environmental change.[16] The Antarctic Treaty System here provides for a regime similar to UNCLOS: while guaranteeing the freedom of scientific investigation, "scientific observations and results from Antarctica shall be exchanged and made freely available."[17] It is not clear how much this provision encompasses the (commercial) utilization of Antarctic genetic resources. Yet a successful conclusion of an ABS regime for MGR could potentially rekindle the debate on creating a legal framework for benefit-sharing in Antarctica.

Both UNCLOS and the Antarctic Treaty address genetic resources from the perspective of scientific research. The Nagoya Protocol also requires parties to create "conditions to promote and encourage research [...] including through simplified measures on access for non-commercial research purposes."[18] The question of how to differentiate between non-commercial research and commercial utilization is critical. It is important to recall that ABS is not (or not supposed to be) about monetary transfers from rich to poor countries only. If we take the normative objections to the ways in which genetic resources are presently being utilized seriously, then ABS is ultimately about finding the proper balance between the public interest and commercial interests. The public interest entails fostering scientific advances in areas that do not immediately lead to profitable inventions. The Earth's genetic heritage is a treasure chest full of opportunities that science simply has to open, and that may improve the lives of subsequent generations in countless and presently probably quite unimaginable ways. The use of this wealth should not be foreclosed by national egoisms. Both UNCLOS and the Nagoya Protocol allow for the differential treatment of non-commercial research, yet the danger is that governments are looking simply at the short-term gains to be had from commercialization and benefit-sharing. While under-regulation will fail to resolve the issue of inequity in the biotechnological utilization of genetic resources, over-regulation will hamper the production of new scientific knowledge—a global public good *par excellence*. It is thus crucial to find a proper balance that, in turn, would require governments to take a broader view of the issues at stake and go beyond reiterating those positions that have been irreconcilable for over 30 years. A stronger emphasis on the sharing of scientific knowledge would require developing countries to move away from their parochial fixation on the sharing of monetary benefits, itself spurred by what is arguably a dramatic overestimation of the commercial value of genetic resources.

Similarly, it would require industrialized countries to take more seriously their existing obligations regarding the sharing of the results of marine scientific research and, where relevant, research in Antarctica. It would also require them to take a long, hard look at intellectual property rights and the balance between public and private interests. The transformation of global plant breeding towards a system in which an ever-smaller number of companies uses the ever-expanding protection of intellectual property to maximize their commercial profits drastically highlights the risks of failing to maintain a proper balance. The public sector has showed an impressive capacity to provide tangible solutions to the world's food crises in the middle of the 20th century, whereas contemporary agribusiness has so far failed to deliver on its lofty promises of sustainability, food security, empowerment of rural communities, conservation of biological diversity and whatnot. Under a market-radical ideology, public plant breeding was dismantled throughout the 1980s and 1990s, yet the alternative paradigm that has taken its place appears to be not part of the solution but part of the problem.[19]

Re-orienting ABS governance towards the fostering of science and the sharing of knowledge, including by limiting unreasonable restrictions from intellectual property rights, would be politically challenging—which is probably an understatement if there ever was one. Yet the fruits we still reap today from open science, the free exchange of knowledge and human creativity in the service of the public interest, show how benefits may accrue over the long term, which may be difficult to foresee over the short one. The political decisions that will be remembered will not be those with which governments sought to maximize their individual gains, but those that made a sizeable contribution to health, sustainable development, efficient resource use, the rights of indigenous peoples and farming communities or the conservation of the world's biological heritage.

Notes

1 Wynne (1989).
2 Clapp (1994).
3 Marcoux and Urpelainen (2012).
4 Dill and Kopsick (2014).
5 Biermann et al. (2009); van Asselt (2014).
6 Sell (2010).
7 Helfer (2009); Morin (2009); Muzaka (2010); Rabitz (2014).
8 Keohane and Victor (2011), 8.
9 Selin and Eckley (2003).
10 Lallas (2001); Kohler and Ashton (2010).
11 Green and Bohannan (2006).
12 Gostin et al. (2014).
13 Tvedt and Schei (2014).
14 Dedeurwaerdere (2010).
15 Oberthür and Pozarowska (2013).
16 Chown et al. (2015).
17 Antarctic Treaty, articles 2 and 3(c).
18 Nagoya Protocol, article 8(a).
19 Murphy (2007b).

Bibliography

Primary sources

05Geneva1648. WIPO Standing Committee on the Law of Patents (SCP) Fails (Again) to Adopt a Work Plan on Substantive Patent Law Harmonization, Fate of Harmonization at WIPO in Doubt. US Embassy to the United Nations, Geneva. 5 July 2005. Sensitive.

05Geneva2798. WTO TRIPS Council October 25–26 2005. US Mission United Nations (Geneva). 16 November 2005. Unclassified.

07Brussels1459. Special Representative John Lange and EU Officials Meet on Avian and Pandemic Influenza. US Mission to European Union. 2 May 2007. Sensitive.

07Jakarta310. Avian Influenza (AI) Sample Sharing Update. US Embassy Jakarta. 6 February 2007. Sensitive.

08Geneva1112. WHO: December 8–13 2008 WHO Meeting on Influenza Sample/Benefits Sharing Makes Substantial Progress, but Issues Remain. US Embassy to the United Nations, Geneva. 22 December 2008. Sensitive.

08Jakarta806. NAMRU-2 – Medical Research Unit's Days May Be Numbered. US Embassy Jakarta. 22 April 2008. Classified.

A60/Inf.Doc/1. Avian and Pandemic Influenza. Best Practices for Sharing Influenza Viruses and Sequence Data. 22 March 2007.

A/61/65. Report of the Ad Hoc Open-Ended Working Group to Study Issues Relating to the Conservation and Sustainable Use of Marine Biological Diversity Beyond Areas of National Jurisdiction. UN General Assembly, 61st session. 20 March 2006.

A62/5. Pandemic Influenza Preparedness: Sharing of Influenza Viruses and Access to Vaccines and Other Benefits. Report by the Secretariat. 30 April 2009.

A/63/79, Annex. Joint Statement of the Co-Chairpersons of the Ad Hoc Open-Ended Informal Working Group to Study Issues Relating to the Conservation and Sustainable Use of Marine Biological Diversity Beyond Areas of National Jurisdiction. UN General Assembly, 63rd session. 16 May 2008.

A64/8. Pandemic Influenza Preparedness: Sharing of Influenza Viruses and Access to Vaccines and Other Benefits. Report by the Open-Ended Working Group of Member States on Pandemic Influenza Preparedness: Sharing of Influenza Viruses and Access to Vaccines and Other Benefits. 5 May 2011.

A/66/119, Annex. Recommendations of the Ad Hoc Open-Ended Informal Working Group to Study Issues Relating to the Conservation and Sustainable Use of Marine Biological Diversity Beyond Areas of National Jurisdiction and Co-Chairs' Summary of Discussion. 30 June 2011.

A/67/95, Annex. Report of the Ad Hoc Open-Ended Informal Working Group to Study Issues Relating to the Conservation and Sustainable Use of Marine Biological Diversity Beyond Areas of National Jurisdiction and Co-chairs' Summary of Discussions. 8 June 2012.

African Group (2012) The Need for and Modalities of a Global Multilateral Benefit-Sharing Mechanism (Article 10). Second meeting of the Intergovernmental Committee for the Nagoya Protocol. New 2 to 6 July 2012.

A/PIP/IGM/4. Sharing of Influenza Viruses and Access to Vacccines and Other Benefits: Interdisciplinary Working Group on Pandemic Influenza Preparedness. Report by the Director-General. 9 October 2007.

A/PIP/IGM/5. Sharing of Influenza Viruses and Access to Vacccines and Other Benefits: Interdisciplinary Working Group on Pandemic Influenza Preparedness. 19 November 2007.

A/PIP/IGM/7. Sharing of Influenza Viruses and Access to Vacccines and Other Benefits: Interdisciplinary Working Group on Pandemic Influenza Preparedness. 4 January 2008.

A/PIP/IGM/8. Reports of the Director-General. Establishment of the Advisory Mechanism. 27 November 2008.

A/PIP/IGM/WG/5. Open-Ended Working Group: Report on Progress to Date. 4 April 2008.

A/PIP/IGM/WG/6. Chair's Text. Pandemic Influenza Preparedness Framework for the Sharing of Influenza Viruses and Access to Vaccines and Other Benefits. Draft. 28 September 2008.

A/RES/54/33. Results of the Review by the Commission on Sustainable Development on the Sectoral Theme of "Oceans and Seas": International Coordination and Cooperation. Resolution adopted by the General Assembly, 18 January 2000.

A/RES/57/141. Regular Process for Global Reporting and Assessment of the State of the Marine Environment, Including Socio-Economic Aspects. Resolution adopted by the General Assembly, 21 February 2003.

A/RES/68/70. Oceans and the Law of the Sea. Resolution adopted by the General Assembly, 9 December 2013.

A/RES/69/292. Development of an International Legally Binding Instrument under the United Nations Convention on the Law of the Sea on the Conservation and Sustainable Use of Marine Biological Diversity of Areas Beyond National Jurisdiction. Resolution adopted by the General Assembly, 19 June 2015.

Brazil. Statement by H.E. Ambassador Carlos Duarte, Depute Permanent Representative of Brazil to the United Nations. 28 March 2016. First Session of the Preparatory Committee Established by General Assembly Resolution 69/292.

C 2001/PV. FAO Conference, Thirty-first Session. Verbatim Records of Plenary Meetings of the Conference. 2–13 November 2001.

CGRFA/CG-1/99/TXT. Revision of the International Undertaking on Plant Genetic Resources, in Harmony with the Convention on Biological Diversity. 20–24 September 1999.

CGRFA/CG-2/00/TXT. Revision of the International Undertaking on Plant Genetic Resources, in Harmony with the Convention on Biological Diversity. 3–7 April 2000.

CGRFA/CG-3/00/TXT. Revision of the International Undertaking on Plant Genetic Resources, in Harmony with the Convention on Biological Diversity. 26–31 August 2000.

CGRFA/CG-6/01/2. Composite Draft Text of the International Undertaking on Plant Genetic Resources. 22–28 April 2001.

CGRFA/IUND/CNT. Consolidated Negotiating Text Resulting from the Deliberations During the Fourth Extraordinary Session of the Commission on Genetic Resources for Food and Agriculture. No date.

CGRFA-Ex3/96/3. Report by the Chairman of the Eleventh Session of the Working Group of the Commission on Genetic Resources for Food and Agriculture. 9–13 December 1996.

CGRFA-Ex3/96/Rep. Report of the Commission on Genetic Resources for Food and Agriculture. 9–13 December 1996.

CGRFA/Ex-6/01/3. Composite Draft Text of the International Undertaking on Plant Genetic Resources. 24–30 June 2001.

CGRFA-8/99/13, Annex. Composite Draft Text of the International Undertaking on Plant Genetic Resources Incorporating the Chairman's Elements. 19–23 April 1999.

CGRFA-8/99/Inf.9. Background Documentation Provided by the International Association of Plant Breeders for the Protection of Plant Varieties (ASSINSEL). 19–23 April 1999.

COM(2012) 494 Final. Blue Growth. Opportunities for Marine and Maritime Sustainable Growth. Communication from the Commission to the European Parliament, the Council, the European Economic and Social Committee and the Committee of the Regions.

Decision II/10. Conservation and Sustainable Use of Marine and Coastal Biological Diversity. Jakarta, 6 to 17 November 1995.

Decision II/11. Access to Genetic Resources. Jakarta, 6–17 November 1995.

Decision II/15. FAO Global System for the Conservation and Utilization of Plant Genetic Resources for Food and Agriculture. Jakarta, 6–17 November 1995.

Decision IV/8. Access and Benefit-Sharing. Fourth Meeting of the Conference of the Parties to the Convention on Biological Diversity. 4–15 May 1998.

Decision V/26. Access to Genetic Resources. Fifth Meeting of the Conference of the Parties to the Convention on Biological Diversity. 15 to 26 May 2000.

Decision VII/28. Protected Areas. Kuala Lumpur, 9 to 20 February 2004.

Decision VIII/21. Marine and Coastal Biological Diversity: Conservation and Sustainable Use of Deep Seabed Genetic Resources Beyond the Limits of National Jurisdiction. Curitiba, 20 to 31 March 2006.

Decision VIII/4. Access and Benefit-Sharing. Eighth Meeting of the Conference of the Parties to the Convention on Biological Diversity. 20 to 31 March 2006.

Decision VIII/24. Protected Areas. Curitiba, 20 to 31 March 2006.

Decision XI/1. Status of the Nagoya Protocol on Access to Genetic Resources and the Fair and Equitable Sharing of Benefits Arising from their Utilization and Related Developments. Hyderabad, 8 to 19 October 2012.

Decision XI/17, Annex. Summary Reports on the Description of Areas Meeting the Scientific Criteria for Ecologically or Biologically Sensitive Marine Areas. Hyderabad, 8 to 19 October 2012.

Decision XI/20. Climate-Related Geoengineering. Decision adopted by the Conference of the Parties to the Convention on Biological Diversity at its Eleventh Meeting. Hyderabad, 8 to 19 October 2012.

EB/120/15. Avian and Pandemic Influenza: Developments, Response and Follow-up, and Application of the International Health Regulations (2005). Report by the Secretariat. 14 December 2006.

EB122/2008/REC/2. Executive Board, 122nd Session. Summary Records. 21–25 January 2008.

EB124/4 Add.1. Pandemic Influenza Preparedness: Sharing of Influenza Viruses and Access to Vaccines and Other Benefits. Resumed Intergovernmental Meeting. Report by the Director-General. 19 January 2009.

EB126/4. Pandemic Influenza Preparedness: Sharing of Influenza Viruses and Access to Vaccines and Other Benefits. Outcome of the Process to Finalize Remaining Elements under the Pandemic Influenza Preparedness Framework for the Sharing of Influenza Viruses and Access to Vaccines and Other Benefits. Report by the Secretariat. 10 December 2009.

EB128/4. Pandemic Influenza Preparedness: Sharing of Influenza Viruses and Access to Vaccines and Other Benefits. Report by the Director-General. 12 January 2011.

G77/China. Marine Genetic Resources Including Questions on the Sharing of Benefits. Group of 77's and China's Intervention. First session of the Preparatory Committee Established by General Assembly Resolution 69/292. No date. 2016.

India. Statement of India at the UN General Assembly Established Preparatory Committee on Development of an International Legally Binding Instrument under the UNCLOS on the Conservation and Sustainable Use of Marine Biological Diversity of Areas Beyond National Jurisdiction [sic]. No date. First Session of the Preparatory Committee Established by General Assembly Resolution 69/292. 2016.

IP/C/W/254. Review of the Provisions of Article 27.3(b) of the TRIPS Agreement. Communication from the European Communities and their Member States. 13 June 2001.

IP/C/W/368. The Relationship between the TRIPS Agreement and the Convention on Biological Diversity. Summary of Issues Raised and Points Made. 8 August 2002.

IP/C/W/383. Review of Article 27.3(b) of the TRIPS Agreement, and the Relationship between the TRIPS Agreement and the Convention on Biological Diversity (CBD) and the Protection of Traditional Knowledge and Folklore. A Concept Paper. Communication from the European Communities and their Member States. 17 October 2002.

IP/C/W/404. Taking Forward the Review of Article 27.3(b) of the TRIPS Agreement. Joint Communication from the African Group. 26 June 2003.

IP/C/W/447. Article 27.3(b), Relationship between the TRIPS Agreement and the CBD and Protection of Traditional Knowledge and Folklore. Communication from Peru. 8 June 2005.

IP/C/W/473. Amending the TRIPS Agreement to Introduce an Obligation to Disclose the Origin of Genetic Resources and Traditional Knowledge in Patent Applications. Communication from Norway. 14 June 2006.

Japan. Consideration of the Scope of an International Legally Binding Instrument and its Relationship with Other Instruments. Statement of the Delegation of Japan. First session of the Preparatory Committee Established by General Assembly Resolution 69/292. Agenda item 7. 28 March 2016.

New Zealand. Scope of an International Legally Binding Instrument and its Relationship with other Instruments. New Zealand. First session of the Preparatory Committee Established by General Assembly Resolution 69/292. Agenda item 7. 29 March 2016.

Nairobi Final Act of the Conference for the Adoption of the Agreed Text of the Convention on Biological Diversity. 20–21 May 1992.

PCT/R/WG/4/13. Proposals by Switzerland Regarding the Declaration of the Source of Genetic Resources and Traditional Knowledge in Patent Applications. Working Group on Reform of the Patent Cooperation Treaty (PCT), fourth session. 19 to 23 May 2003.

PCT/R/WG/5/13. Working Group on Reform of the Patent Cooperation Treaty (PCT). Summary of the Session. Fifth session. 17 to 21 November 2003.

PCT/R/WG/6/11. Additional Comments by Switzerland on its Proposals Regarding the Declaration of the Source of Genetic Resources and Traditional Knowledge in Patent Applications. Working Group on Reform of the Patent Cooperation Treaty (PCT), sixth session. 3 to 7 May 2004.

PCT/R/WG/6/12. Working Group on Reform of the Patent Cooperation Treaty (PCT). Summary of the Session. Sixth session. 3 to 7 May 2004.

Recommendation VIII/3. Marine and Coastal Biodiversity: Review, Further Elaboration and Refinement of the Programme of Work. Subsidiary Body on Scientific, Technical and Technological Advice. Montreal, 10 to 14 March 2003.

Resolution 3/91. Annex 3 to the International Undertaking on Plant Genetic Resources. 26th Session of the FAO Conference. 9–27 November 1991.

Resolution 7/93. Revision of the International Undertaking on Plant Genetic Resources. 27th Session of the FAO Conference. 6–24 November 1993.

SCP/3/10. Protection of Biological and Genetic Resources. Proposal by the Delegation of Colombia. Standing Committee on the Law of Patents. Third Session. 6 to 14 September 1999.

SCP/4/2. Suggestions for the Further Development of International Patent Law. Standing Committee on the Law of Patents. Fourth session. 6 to 10 November 2000.

SCP/7/8. Report. Standing Committee on the Law of Patents. Seventh session. 6 to 10 May 2002.

SCP/9/8. Report. Standing Committee on the Law of Patents. Ninth session. 12 to 16 May 2003.

SCP/10/11. Report. Standing Committee on the Law of Patents. Tenth session. 10 to 14 May 2004.

Switzerland. Prepcom: Swiss Considerations on the Integration of the Marine Genetic Resources Component in the International Legally Binding Instrument. Second session of the Preparatory Committee Established by General Assembly Resolution 69/292. Agenda item 6. 29 August 2016.

TN/C/W/52. Draft Modalities for TRIPS Related Issues. 19 July 2008.

UNEP/CBD/ABS/7/INF/1/Add. 1. Compilation of Submissions by Parties, Governments, International Organizations, Indigenous and Local Communities and Relevant Stakeholders in Respect of the Main Components of the International Regime on Access and Benefit-Sharing Listed in Decision IX/12, Annex I (Addendum). 25 March 2009.

UNEP/CBD/WG-ABS/5/INF/2. Compilation of Submissions by Parties on Experiences in Developing and Implementing Article 15 of the Convention at the National Level and Measures Taken to Support Compliance with Prior Informed Consent and Mutually Agreed Terms. 20 July 2007.

UNEP/CBD/WG-ABS/6/INF/3. Compilation of Submissions Provided by Parties, Governments, Indigenous and Local Communities and Stakeholders on Concrete Options on Substantive Items on the Agenda of the Fifth and Sixth Meeting of the Ad Hoc Open-Ended Working Group on Access and Benefit-Sharing. 13 December 2007.

UNEP/CBD/WG-ABS/7/2. Report of the Meeting of the Group of Legal and Technical Experts on Concepts, Terms, Working Definitions and Sectoral Approaches. 12 December 2008.

UNEP/CBD/WG-ABS/7/8. Report of the Seventh Meeting of the Ad Hoc Open-Ended Working Group on Access and Benefit-Sharing. 5 May 2009.

WHA60/2007/REC/3. Summary Records of Committees. Sixtieth World Health Assembly. 14 to 23 May 2007.

WIPO/GRTKF/IC/8/11. Disclosure of Origin or Source of Genetic Resources and Associated Traditional Knowledge in Patent Applications. Intergovernmental Committee on Intellectual Property Rights, Traditional Knowledge and Folklore. Eighth Session. 6 to 10 June 2005.

WIPO/GRTKF/IC/18/9. Draft Objectives and Principles Relating to Intellectual Property and Genetic Resources Prepared at IWG 3. Intergovernmental Committee on Intellectual Property Rights, Traditional Knowledge and Folklore. Eighteenth Session. 9 to 13 May 2011.

WIPO/GRTKF/IC/19/6. The Protection of Traditional Knowledge: Draft Objectives and Principles. Intergovernmental Committee on Intellectual Property Rights, Traditional Knowledge and Folklore. Tenth Session. 30 November to 8 December 2006.

WO/GA/26/6. Matters Concerning Intellectual Property and Genetic Resources, Traditional Knowledge and Folklore. Document Prepared by the Secretariat. WIPO General Assembly, Twenty-Sixth (12th extraordinary) Session, 25 September to 3 October 2000.

WT/MIN(01)/DEC/1. Ministerial Declaration. Doha WTO Ministerial, 20 November 2001.

WT/MIN(11)/11. Eighth Ministerial Conference of the World Trade Organization. Chairman's Concluding Statement. 17 December 2011.

Secondary sources

Abbott, Frederick M. (2010) *An International Legal Framework for the Sharing of Pathogens: Issues and Challenges*. Geneva: International Center for Trade and Sustainable Development.

Abbott, Frederick M., and Graham Dukes (2009) *Global Pharmaceutical Policy. Ensuring Medicines for Tomorrow's World*. Cheltenham: Edward Elgar.

Abbott, Kenneth W., Robert O. Keohane, Andrew Moravcsik, Anne-Marie Slaughter, and Duncan Snidal (2000) The Concept of Legalization. *International Organization* 54(3): 401–419.

Ackrill, Robert, and Adrian Kay (2014) *The Growth of Biofuels in the 21st Century: Policy Drivers and Market Challenges*. Basingstoke: Palgrave.

Alter, Karen J., and Sophie Meunier (2009) The Politics of International Regime Complexity. *Perspectives on Politics* 7(1): 13–24.

Andersen, Regine (2005) *The History of Farmers' Rights: A Guide to Central Documents and Literature*. FNI Report 8/2005. Oslo: Fridtjof Nansen Institute.

Andersen, Regine (2008) *Governing Agrobiodiversity. Plant Genetics and Developing Countries*. Aldershot: Ashgate.

Arnaud-Haond, Sophie, Jesús M. Arrieta, and Carlos M. Duarte (2011) Marine Biodiversity and Gene Patents. *Science* 331: 1521–1522.

Atkinson, Giles, Ian J. Bateman, and Susanna Mourato (2014) Valuing Ecosystem Services and Biodiversity. In: Dieter Helm and Cameron Hepburn (eds.), *Nature in the Balance. The Economics of Biodiversity*. Oxford: Oxford University Press.

Aubertin, Catherine, and Geoffroy Filoche (2011) The Nagoya Protocol on the Use of Genetic Resources: One Embodiment of an Endless Discussion. *Sustentabilidade em Debate* 2(1): 51–64.

Avian Influenza Action Group (2008) *Avian and Pandemic Influenza. The Global Response*. Washington, DC: US Department of State.

Axelrod, Mark (2011) Savings Clauses and the "Chilling Effect": Regime Interplay as Constraints on International Governance. In: Sebastian Oberthür and Olav S. Stokke (eds.), *Managing Institutional Complexity: Regime Interplay and Global Environmental Change*. Cambridge, MA: MIT Press, 87–114.

Azar, Christian, Kristian Lindgren, Michael Obersteiner, Keywan Riahi, Detlef P. van Vuuren, K. Michel G. J. den Elzen, Kenneth Möllersten, and Eric D. Larson (2010) The Feasibility of Low CO_2 Concentration Targets and the Role of Bio-energy with Carbon Capture and Storage (BECCS). *Climatic Change* 100(1): 195–202.

Bagley, Margo A., and Arti K. Rai (2013) The Nagoya Protocol and Synthetic Biology Research: A Look at the Potential Impacts. *Synbio* 6 / November 2013. Woodrow Wilson Center.

Baram, Michael, and Mathilde Bourrier (2011) Governing Risk in GM Agriculture: An Introduction. In: Michael Baram and Mathilde Bourrier (eds.), *Governing Risk in GM Agriculture*. Cambridge: Cambridge University Press, 1–12.

Barkin, Samuel J., and Elizabeth R. DeSombre (2013) *Saving Global Fisheries. Reducing Fishing Capacity to Promote Sustainability*. Cambridge, MA / London: MIT Press.

Barrett, Scott (2007) *Why Cooperate? The Incentive to Supply Global Public Goods*. Oxford: Oxford University Press.

Barrows, Geoffrey, Steven Sexton, and David Zilberman (2014) Agricultural Biotechnology: The Promise and Prospects of Genetically Modified Crops. *Journal of Economic Perspectives* 28(1): 99–120.

Barthlott, Willhelm, Jens Mutke, Daud Rafiqpoor, Georld Kier, and Holger Kreft (2005) Global Centers of Vascular Plant Diversity. *Nova Acta Leopoldina* 92(342): 61–83.

Batta Bjørnstad, Svanhild-Isabelle (2004) *Breakthrough for 'the South'? An Analysis of the Recognition of Farmers' Rights in the International Treaty on Plant Genetic Resources for Food and Agriculture*. FNI Report 13/2004. Oslo: Fridtjof Nansen Institute.

Bennett, Andrew, and Jeffrey T. Checkel (2015) Process Tracing: From Philosophical Roots to Best Practices. In: Andrew Bennet and Jeffrey T. Checkel (eds.), *Process Tracing. From Metaphor to Analytical Tool*. Cambridge: Cambridge University Press.

Bernauer, Thomas, Anna Kalbhenn, Vally Koubi, and Gabriele Spilker (2013) Is There a "Depth versus Participation" Dilemma in International Cooperation? *Review of International Organizations* 8(4): 477–497.

Betts, Alexander (2009) Institutional Proliferation and the Global Refugee Regime. *Perspectives on Politics* 7(1): 53–58.

Beyer, Jürgen (2010) The Same or Not the Same – On the Variety of Mechanisms of Path Dependence. *International Journal of Social Sciences* 5(1): 1–11.

Biermann, Frank, Phillip Pattberg, Harro van Asselt, and Fariborz Zelli (2009) The Fragmentation of Global Governance Architectures: A Framework for Analysis. *Global Environmental Politics* 9(4): 14–40.

Blasiak, Robert, and Nobuyuki Yagi (2016) Shaping an International Agreement on Marine Biodiversity Beyond Areas of National Jurisdiction: Lessons from High Seas Fisheries. *Marine Policy* 71: 210–216.

Bordwin, Harold J. (1985) The Legal and Political Implications of the International Undertaking on Plant Genetic Resources. *Ecology Law Quarterly* 12(4): 1053–1069.

Borowiak, Craig (2004) Farmers' Rights: Intellectual Property Regimes and the Struggle over Seeds. *Politics & Society* 32(4): 511–543.

Brand, Ulrich, Christoph Görg, Joachim Hirsch, and Markus Wissen (2008) *Conflicts in Environmental Regulation and the Internationalisation of the State: Contested Terrains*. Oxon: Routledge.

Brennan, Margaret F., Carl E. Pray, and Ann Courtmanche (2000) Impact of Industry Concentration on Innovation in the U.S. Plant Biotech Industry. In: William H. Lesser (ed.), *Transitions in Agbiotech: Economics of Strategy and Policy*. Connecticut: Food Marketing Policy Center.

Broggiato, Arianna (2011) Marine Genetic Resources beyond National Jurisdiction. Coordination and Harmonization of Governance Regimes. *Environmental Policy and Law* 41(1): 35–43.

Broggiato, Arianna, Sophie Arnaud-Haond, Claudio Chiarolla, and Thomas Greiber (2014) Fair and Equitable Sharing of Benefits from the Utilization of Marine Genetic Resources in Areas Beyond National Jurisdiction: Bridging the Gaps between Science and Policy. *Marine Policy* 49: 176–185.

Brookes, Graham, and Peter Barfoot (2008) Global Impact of Biotech Crops: Socio-economic and Environmental Effects, 1996–2006. *AgBioForum* 11(1): 21–38.

Buck, Matthias (2008) The Main Results of the Ninth Conference of the Parties to the UN Convention on Biological Diversity. *Journal for European Environmental Planning Law* 5(3): 249–261.

Buck, Matthias, and Clare Hamilton (2011) The Nagoya Protocol on Access to Genetic Resources and the Fair and Equitable Sharing of Benefits Arising from their Utilization to the Convention on Biological Diversity. *Review of European Community & International Environmental Law* 20(1): 47–61.

Carr, Jonathan (2008) Agreements that Divide: TRIPS vs. CBD and proposals for Mandatory Disclosure of Source and Origin of Genetic Resources in Patent Applications. *Journal of Transnational Law & Policy* 18: 131–154.

de la Calle, Fernando (2009) Marine Genetic Resources. A Source of New Drugs. The Experience of the Biotechnology Sector. *The International Journal of Marine and Coastal Law* 24: 209–220.

Carvalho, Loic, G., and Leonel Pereira (2015) Review of Marine Algae as Source of Bioactive Metabolites: A Marine Biotechnology Approach. In: Leonel Pereira and João M. Neto (eds.), *Marine Algae. Biodiversity, Taxonomy, Environmental Assessment, and Biotechnology*. Boca Raton / New York / London: CRC Press.

de Carvalho, Nuno Pires (2005) From the Shaman's Hut to the Patent Office: In Search of a TRIPS-Consistent Requirement to Disclose the Origin of Genetic Resources and Prior Informed Consent. *Washington University Journal of Law & Policy* 17: 111–186.

Capano, Giliberto (2009) Understanding Policy Change as an Epistemological and Theoretical Problem. *Journal of Comparative Policy Analysis* 11(1): 7–31.

CBD (2015) *Synthetic Biology*. CBD Technical Series No. 82. Montreal: Secretariat of the Convention on Biological Diversity.

Ceccarelli, Salvatore (2009) Main Stages of a Plant Breeding Programme. In: Salvatore Ceccarelli, Elcio Guimarães, and Eva Weltzien (eds.), *Plant Breeding and Farmer Participation*. Rome: Food and Agriculture Organization, 63–74.

Cefis, Elena, Matteo Ciccarelli, and Luigi Orsenigo (2006) Heterogeneity and Firm Growth in the Pharmaceutical Industry. In: Mariana Mazzucato and Giovanni Dosi (eds.), *Knowledge Accumulation and Industry Evolution. The Case of Pharma-Biotech*. Cambridge: Cambridge University Press, 163–207.

Chandler, Alfred D. (2005) *Shaping the Industrial Century. The Remarkable Story of the Evolution of the Modern Chemical and Pharmaceutical Industries*. Cambridge, MA / London: Harvard University Press.

Chaturvedi, Sachin (2009) Agricultural Biotechnology and Trends in the Intellectual Property Rights Regime: Emerging Challenges for Developing Countries. In: David

Castle (ed.), *The Role of Intellectual Property Rights in Biotechnology Innovation.* Cheltenham: Edward Elgar, 369–391.

Chayes, Abraham, and Antonia H. Chayes (1993) On Compliance. *International Organization* 47(2): 175–205.

Chiarolla, Claudio (2008) Plant Patenting, Benefit Sharing and the Law Applicable to the Food and Agriculture Organisation Standard Material Transfer Agreement. *Journal of World Intellectual Property* 11(1): 1–28.

Chiarolla, Claudio, and Hope Shand (2013) An Assessment of Private *Ex Situ* Collections: The Private Sector's Participation in the Multilateral System of the FAO International Treaty on Plant Genetic Resources for Food and Agriculture. Berne Declaration/ Development Fund.

Chown, Steven L., Andrew Clarke, Ceridwen I. Fraser, Craig Carry, Katherine L. Moon, and Melodie A. McGeoch (2015) The Changing Form of Antarctic Biodiversity. *Nature* 522: 431–438.

Cicin-Sain, Biliana, Miriam Balgos, Joseph Appiott, Kateryna Wowk, and Gwénaëlle Hamon (2011) *Oceans at Rio +20: How Well Are We Doing in Meeting the Commitments from the 1992 Earth Summit and the 2002 World Summit on Sustainable Development?* Newark: Global Ocean Forum.

Clapp, Jennifer (1994) The Toxic Waste Trade with Less-Industrialised Countries: Economic Linkages and Political Alliances. *Third World Quarterly* 15(3): 505–518.

Cockburn, Iain M. (2004) The Changing Structure of the Pharmaceutical Industry. *Health Affairs* 23(1): 10–22.

Colgan Jeff D., Robert O. Keohane, and Thijs van de Graaf (2012) Punctuated Equilibrium in the Energy Regime Complex. *The Review of International Organizations* 7(2):117–143.

Collins, John, Ned Hall, and L. A. Paul (2004) Counterfactuals and Causation: History, Problems, and Prospects. In: John Collins, Ned Hall, and L. A. Paul (eds.), *Causation and Counterfactuals.* Cambridge, MA / London: MIT Press, 1–57.

Coolsaet, Brendan (2015) Conclusion. Comparing Access and Benefit-Sharing in Europe. In: Brendan Coolsaet, Fulya Batur, Arianna Broggiato, John Pitseys, and Tom Dedeurwaerdere (eds.), *Implementing the Nagoya Protocol. Comparing Access and Benefit-Sharing Regimes in Europe.* Leiden / Boston: Brill Nijhoff, 363–402.

Coolsaet, Brendan, Fulya Batur, Arianna Broggiato, John Pitseys, Tom Dedeurwaerdere (eds.) (2015) *Implementing the Nagoya Protocol: Comparing Access and Benefit-Sharing Regimes in Europe.* Leiden / Boston: Brill Nijhoff.

Coolsaet, Brendan, and John Pitseys (2015) Fair and Equitable Negotiations? African Influence and the International Access and Benefit-Sharing Regime. *Global Environmental Politics* 15(2): 38–56.

Cooper, David H. (1993) The International Undertaking on Plant Genetic Resources. *Review of European, Comparative and International Environmental Law* 2(2): 158–166.

Cooper, David H. (2002) The International Treaty on Plant Genetic Resources for Food and Agriculture. *Review of European, Comparative and International Environmental Law* 11(1): 1–16.

Cornes, Richard, and Todd Sandler (1996) *The Theory of Externalities, Public Goods, and Club Goods.* Cambridge: Cambridge University Press.

Correa, Carlos M. (2006) Considerations on the Standard Material Transfer Agreement under the FAO Treaty on Plant Genetic Resources for Food and Agriculture. *Journal of World Intellectual Property* 9(2): 137–165.

Coupe, Stuart, and Roger Lewins (2007) *Negotiating the Seed Treaty.* Rugby: Practical Action Publishing.

Crowder, Larry, and Elliott Norse (2008) Essential Ecological Insights for Marine Ecosystem-Based Management and Marine Spatial Planning. *Marine Policy* 32(5): 772–778.

Curci, Jonathan (2010) *The Protection of Biodiversity and Traditional Knowledge in International Law of Intellectual Property.* Cambridge: Cambridge University Press.

Davis, Christina L. (2009) Overlapping Institutions in Trade Policy. *Perspectives on Politics* 7(1): 25–31.

Dedeurwaerdere, Tom (2010) Global Microbial Commons: Institutional Challenges for the Global Exchange and Distribution of Microorganisms in the Life Sciences. *Research in Microbiology* 161(6): 414–421.

Dhar, Biswajit (2002) *Sui Generis Systems for Plant Variety Protection. Options under TRIPS.* Geneva: Quaker United Nations Office.

Diamond, Jared (2005) *Guns, Germs and Steel.* London: Vintage.

Dill, David C., and Deborah A. Kopsick (2014) Improving Cooperation between Customs and Environmental Agencies to Prevent Illegal Transboundary Shipments of Hazardous Waste. *World Customs Journal* 8(2): 47–61.

Doherty, Peter C. (2013) *Pandemics. What Everyone Needs to Know.* New York: Oxford University Press.

Dosi, Giovanni, and Mariana Mazzucato (2006) Introduction. In: Mariana Mazzucato and Giovanni Dosi (eds.), *Knowledge Accumulation and Industry Evolution. The Case of Pharma-Biotech.* Cambridge: Cambridge University Press, 1–18.

Downs, George W., David M. Rocke, and Peter N. Barsoom (1996) Is the Good News about Compliance Good News about Cooperation? *International Organization* 50(3): 379–406.

Drahos, Peter (1996) *A Philosophy of Intellectual Property.* Aldershot: Ashgate.

Drankier, Petra (2012) Marine Protected Areas in Areas Beyond National Jurisdiction. *The International Journal of Marine and Coastal Law* 27: 291–350.

Drankier, Petra, Alex G. Oude Elferink, Bert Visser, and Tamara Takács (2012) Marine Genetic Resources in Areas Beyond National Jurisdiction: Access and Benefit-Sharing. *The International Journal of Marine and Coastal Law* 27: 375–433.

Drezner, Daniel W. (2007) *All Politics Is Global: Explaining International Regulatory Regimes.* Princeton / Oxford: Princeton University Press.

Drezner, Daniel W. (2009) The Power and Peril of International Regime Complexity. *Perspectives on Politics* 7(1): 65–70.

Druel, Elizabeth (2011) Marine Protected Areas in Areas Beyond National Jurisdiction: The State of Play. *IDDRI Working Paper* 7/11. Paris: Institut du développement durable et des relations internationales.

Druel, Elizabeth, and Kristina M. Gjerde (2014) Sustaining Marine Life Beyond Boundaries: Options for an Implementing Agreement for Marine Biodiversity Beyond National Jurisdiction under the United Nations Convention on the Law of the Sea. *Marine Policy* 49: 90–97.

Dutfield, Graham (2004) *Intellectual Property, Biogenetic Resources and Traditional Knowledge.* London: Earthscan.

Dutfield, Graham (2011) *Food, Biological Diversity and Intellectual Property: The Role of the International Union for the Protection of New Varieties of Plants (UPOV).* Geneva: Quaker United Nations Office.

Dutfield, Graham (2014) Traditional Knowledge, Intellectual Property and Pharmaceutical Innovation. What's Left to Discuss? In: Matthew David and Debora Halbert (eds.), *The SAGE Handbook of Intellectual Property*. London: Sage, 649–664.

Ecorys, Deltares, and Oceanic Development (2012) *Blue Growth. Scenarios and Drivers for Sustainable Growth from the Oceans, Seas and Coasts*. Brussels: European Commission.

Egziabher, Tewolde B.G., Elizabeth Matos, and Godfrey Mwila (2011) The African Regional Group: Creating Fair Play between North and South. In: Christine Frison, Francisco López and José T. Esquinas-Alcázar (eds.) *Plant Genetic Resources and Food Security. Stakeholder Perspectives on Plant Genetic Resources for Food and Agriculture*. London / New York: Earthscan, 41–56.

Eicher, Carl K., and Mandivamba Rukuni (2003) *The CGIAR at 31: An Independent Meta-Evaluation of the Consultative Group on International Agricultural Research*. Washington, DC: World Bank.

Elbe, Stefan (2011) Pandemics on the Radar Screen: Health Security, Infectious Disease and the Medicalisation of Insecurity. *Political Studies* 59(4): 848–866.

ENB (1997) *Seventh Session of the Commission on Genetic Resources for Food and Agriculture: 15–23 May 1997*. International Institute for Sustainable Development: *Earth Negotiation Bulletin* 9(68).

ENB (2000a) Third Inter-sessional Contact Group Meeting on the Revision of the International Undertaking on Plant Genetic Resources, in Harmony with the CBD: 26–31 August 2000. *Earth Negotiation Bulletin* 9(161).

ENB (2000b) Fourth Inter-sessional Contact Group Meeting on the Revision of the International Undertaking on Plant Genetic Resources, in Harmony with the CBD: 12–17 November 2000. *Earth Negotiation Bulletin* 9(167).

ENB (2001) Fifth Inter-sessional Contact Group Meeting on the Revision of the International Undertaking on Plant Genetic Resources, in Harmony with the CBD: 5–10 January 2001. *Earth Negotiation Bulletin* 9(180).

ENB (2003) Summary of the Fourth Meeting of the Open-Ended Informal Consultative Process on Oceans and the Law of the Sea: 2–6 June 2003. *Earth Negotiation Bulletin* 25(6).

ENB (2004) Summary of the Fifth Meeting of the Open-Ended Informal Consultative Process on Oceans and the Law of the Sea: 7–11 June 2003. *Earth Negotiation Bulletin* 25(12).

ENB (2005) Summary of the First Meeting of the CBD *Ad Hoc* Open-Eended Working Group on Protected Areas: 13–17 June 2005. *Earth Negotiation Bulletin* 9(326).

ENB (2006a) Summary of the Fourth Meeting of the Working Group on Access and Benefit-Sharing of the Convention on Biological Diversity: 30 January-3 February 2006. *Earth Negotiation Bulletin* 9(344).

ENB (2006b) Summary of the Working Group on Marine Biodiversity Beyond Areas of National Jurisdiction: 13–17 February 2006. *Earth Negotiation Bulletin* 25(25).

ENB (2006c) Summary of the Eighth Conference of the Parties to the Convention on Biological Diversity: 20–31 March 2006. *Earth Negotiation Bulletin* 9(363).

ENB (2007) Summary of the Eighth Meeting of the Open-Ended Informal Consultative Process on Oceans and the Law of the Sea: 25 to 29 June 2007. *Earth Negotiation Bulletin* 25(43).

ENB (2008a) Summary of the Sixth Meeting of the Working Group on Access and Benefit-Sharing of the Convention on Biological Diversity: 21–25 January 2008. *Earth Negotiation Bulletin* 9(426).

ENB (2008b) Summary of the Second Meeting of the Working Group on Marine Biodiversity Beyond Areas of National Jurisdiction: 2 April to 2 May 2008. *Earth Negotiation Bulletin* 25(49).

ENB (2009) Summary of the Seventh Meeting of the Working Group on Access and Benefit-Sharing of the Convention on Biological Diversity: 2–8 April 2009. *Earth Negotiation Bulletin* 9(465).

ENB (2010) Summary of the Resumed Ninth Meeting of the Working Group on Access and Benefit-Sharing of the Convention on Biological Diversity: 10–16 July 2010. *Earth Negotiation Bulletin* 9(527).

ENB (2012) Summary of the Fifth Meeting of the Working Group on Marine Biodiversity Beyond Areas of National Jurisdiction: 7 to 11 May 2012. *Earth Negotiation Bulletin* 25(83).

ENB (2014a) Summary of the Eighth Meeting of the Working Group on Marine Biodiversity Beyond Areas of National Jurisdiction: 16 to 19 June 2014. *Earth Negotiation Bulletin*.

ENB (2014b) Summary of the Third Meeting of the Intergovernmental Committee for the Nagoya Protocol on Access and Benefit-Sharing to the Convention on Biological Diversity: 24–28 February 2014. *Earth Negotiation Bulletin* 9(617).

ENB (2015) Summary of the Ninth Meeting of the Working Group on Marine Biodiversity Beyond Areas of National Jurisdiction: 20 to 23 January 2015. *Earth Negotiation Bulletin* 25(94).

ENB (2016) Summary of the First Session of the Preparatory Committee on Marine Biodiversity of Areas Beyond National Jurisdiction: 28 March-8 April 2016. *Earth Negotiation Bulletin* 25(106).

Epstein, Gerald L. (2012) DNA Shuffling and Directed Evolution. In: Jonathan B. Tucker (ed.), *Innovation, Dual Use, and Security. Managing the Risks of Emerging Biological and Chemical Technologies.* Cambridge, MA / London: MIT Press, 101–115.

Esquinas-Alcázar, José, Angela Hilmi, and Isabel López Noriega (2013) A Brief History of the Negotiations on the International Treaty on Plant Genetic Resources for Food and Agriculture. In: Michael Halewood, Isabel López Noriega and Selim Louafi (eds.), *Crop Genetic Resources as a Global Commons: Challenges in International Law and Governance.* London: Earthscan, 135–149.

ETC Group (2010) *The New Biomasters: Synthetic Biology and the Next Assault on Biodiversity and Livelihoods.* Montreal: Action Group on Erosion, Technology and Concentration.

ETC Group (2013) *Putting the Cartel before the Horse...and Farm, Seeds, Soil, Peasants, etc. Who Will Control Agricultural Inputs, 2013?* Montreal: Action Group on Erosion, Technology and Concentration.

ETC Group (2016) Craig Venter lays an Easter Egg. Six Years in the Making, "Synthia" is Resurrected. Montreal: Action Group on Erosion, Technology and Concentration.

EVM (2004) *Worldwide Major Vaccine Manufacturers in Figures.* Brussels: European Vaccine Manufacturers.

Falcon, Walter P., and Cary Fowler (2002) Carving up the Commons – Emergence of a New International Regime for Germplasm Development and Transfer. *Food Policy* 27(3): 197–222.

Falkner, Robert (2007) The Political Economy of 'Normative Power' Europe: EU Environmental Leadership in International Biotechnology Regulation. *Journal of European Public Policy* 14(4): 507–526.

Falkner, Robert (2009) *Business Power and Conflict in International Environmental Politics.* Basingstoke: Palgrave.

FAO (2010) *The Second Report on the State of the World's Plant Genetic Resources for Food and Agriculture*. Commission on Genetic Resources for Food and Agriculture. Rome: Food and Agriculture Organization.

FAO (2016) *Climate Change and Food Security: Risks and Responses*. Rome: Food and Agriculture Organization.

de la Fayette, Louise A. (2009) A New Regime for the Conservation and Sustainable Use of Marine Biodiversity and Genetic Resources Beyond the Limits of National Jurisdiction. *International Journal of Marine and Coastal Law* 24(2): 221–280.

Fearon, James D. (1991) Counterfactuals and Hypothesis Testing in Political Science. *World Politics* 43(2): 169–195.

Fidler, David P. (2007) Indonesia's Decision to Withhold Influenza Virus Samples from the World Health Organization: Implications for International Law. *American Society of International Law Insights* 11(4).

Filomeno, Felipe A. (2014) *Monsanto and Intellectual Property in South America*. New York: Palgrave.

Finnemore, Martha, and Kathryn Sikking (1998) International Norm Dynamics and Political Change. *International Organization* 52(4): 887–917.

Fioretos, Orfeo (2011) Historical Institutionalism in International Relations. *International Organization* 65(2): 367–399.

Fraleigh, Brad, and Bryan L. Harvey (2011) The North American Group: Globalization that Works. In: Christine Frison, Francisco López and José T. Esquinas-Alcázar (eds.) *Plant Genetic Resources and Food Security. Stakeholder Perspectives on Plant Genetic Resources for Food and Agriculture*. London / New York: Earthscan, 109–120.

Frank, Andre Gunder (1978) *World Accumulation, 1492–1789*. New York: Algora Publishing.

Frankham, Dick, Jon Ballou, and David Briscoe (2004) *A Primer on Conservation Genetics*. Cambridge: Cambridge University Press.

Fukuda-Parr, Sakiko (2007) Emergence and Global Spread of GM Crops: Explaining the Role of Institutional Change. In: Sakiko Fukuda-Parr (ed.), *The Gene Revolution: GM Crops and Unequal Development*. London: Earthscan, 15–35.

Fulton, Murray, and Konstantinos Giannakas (2001) Agricultural Biotechnology and Industry Structure. *AgBioForum* 4(2): 137–151.

Fuss, Sabine, Josep B. Canadell, Glen P. Peters, Massimo Tavoni, Robbie M. Andrew, Philippe Ciais, Robert B. Jackson, Chris D. Jones, Florian Kraxner, Nebosja Nakicenovic, Corinne Le Quéré, Michael R. Raupach, Ayyoob Sharifi, Pete Smith, and Yoshiki Yamagata (2014) Betting on Negative Emissions. *Nature Climate Change* 4(10): 850–853.

Garavaglia, Christian, Franco Malerba, Luigi Orsenigo, and Michele Pezzoni (2012) Technological Regimes and Demand Structure in the Evolution of the Pharmaceutical Industry. *Journal of Evolutionary Economics* 22(4): 677–709.

Gehring, Thomas, and Benjamin Faude (2014) A Theory of Emerging Order within Institutional Complexes: How Competition among Regulatory International Institutions Leads to Institutional Adaptation and Division of Labor. *Review of International Organizations* 9(4): 471–498.

Gehring, Thomas, and Sebastian Oberthür (2002) Interplay. In: Oran R. Young, Leslie A. King, and Heike Schroeder (eds.), *Institutions and Environmental Change: Principal Findings, Applications, and Research Frontiers*. Cambridge, MA: MIT Press.

George, Alexander L., and Andrew Bennett (2005) *Case Studies and Theory Development in the Social Sciences*. Cambridge, MA: Belfer Center.

Gerbasi, Fernando (2011) Overview of Regional Approaches: The Negotiation Process of the International Treaty on Plant Genetic Resources for Food and Agriculture. In: Christine Frison, Francisco López and José T. Esquinas-Alcázar (eds.) *Plant Genetic Resources and Food Security. Stakeholder Perspectives on Plant Genetic Resources for Food and Agriculture*. London / New York: Earthscan, 3–40.

Gilbert, Natasha (2013) A Hard Look at GM Crops. *Nature* 497(7447): 24–26.

Gilligan, Michael J. (2004) Is There a Broader-Deeper Trade-off in International Multilateral Agreements? *International Organization* 58(3): 459–484.

Ginzky, Harald, and Robyn Frost (2014) Marine Geo-Engineering: Legally Binding Regulation under the London Protocol. *Carbon and Climate Law Review* 8(2): 82–96.

Global Health Watch. (2008) *Second Report*. London / New York: Zed Books.

Goertz, Gary, and James Mahoney (2012) *A Tale of Two Cultures. Qualitative and Quantitative Research in the Social Sciences*. Princeton / Oxford: Princeton University Press.

Goldin, Ian (2013) *Divided Nations. Why Global Governance is Failing, and What We Can Do About It*. Oxford: Oxford University Press.

Gomar, José O.V., Lindsay C. Stringer, and Jouni Paavola (2014) Regime Complexes and National Policy Coherence: Experiences in the Biodiversity Cluster. *Global Governance* 20(1): 119–145.

Gostin, Lawrence O. (2014) *Global Health Law*. London / Cambridge, MA: Harvard University Press.

Gostin, Lawrence O., Alexandra Phelan, Michael A. Stoto, John D. Kraemer, and K. Srinath Reddy (2014) Virus Sharing, Genetic Sequencing, and Global Health Security. *Science* 345(6202): 1295–1296.

Grabowski, Henry, and John Vernon (1994) Innovation and Structural Change in Pharmaceuticals and Biotechnology. *Industrial and Corporate Change* 3(2): 435–449.

Graff, Gregory D., Gordon C. Rausser, and Arthur A. Small (2001) Agricultural Biotechnology's Complementary Intellectual Assets. *Institute for Social and Economic Research and Policy*, working paper 01–08.

Green, Jessica, and Brendan J. Bohannan (2006) Spatial Scaling of Microbial Biodiversity. *Trends in Ecology & Evolution* 21(9): 501–507.

Greenpeace (1999) *Centers of Diversity – Global Heritage of Crop Varieties Threatened by Genetic Pollution*. Berlin: Greenpeace.

Greiber, Thomas, Sonia P. Moreno, and 8 others (2012) *An Explanatory Guide to the Nagoya Protocol on Access and Benefit-Sharing*. IUCN Environmental Policy and Law Paper No. 83.

Gupta, Aarti, Till Pistorius, and Marjanneke J. Vijge (2015) Managing Fragmentation in Global Environmental Governance: The REDD+ Partnership as Bridge Organization. *International Environmental Agreements*, doi: 10.1007/s10784-015-9274-9.

Gurib-Fakim, Ameenah (ed.) (2014) *Novel Plant Bioresources. Applications in Food, Medicine and Cosmetics*. Chichester: Wiley Blackwell.

Gusta, Michael, Stuart J. Smyth, Kenneth Belcher, Peter W.B. Phillips, and David Castle (2011) Economic Benefits of Genetically Modified Herbicide-tolerant Canola for Producers. *AgBioForum* 14(1): 1–13.

Guzzini, Stefano (1993) Structural Power: The Limits of Neorealist Analysis. *International Organization* 47(3): 443–478.

Haas, Ernst B. (1980) Why Collaborate? Issue-Linkage and International Regimes. *World Politics* 32(3): 357–405.

Halewood, Michael, Raj Sood, Ruaraidh Sackville Hamilton, Ahmed Amri, Ines Van den Houwe, Nicolas Roux, Dominique Dumet, Jean Hanson, Hari D. Upadhyaya,

Alexandra Jorge, and David Tay (2013) Changing Rates of Acquisition of Plant Genetic Resources by International Gene Banks: Setting the Scene to Monitor an Impact of the International Treaty. In: Michael Halewood, Isabel López Noriega and Selim Louafi (eds.), *Crop Genetic Resources as a Global Commons: Challenges in International Law and Governance*. London: Earthscan, 99–131.

Halford, Nigel G. (2012) *Genetically Modified Crops*. London: Imperial College Press.

Hameiri, Shahar (2014) Avian Influenza, "viral sovereignty", and the Politics of Health Security in Indonesia. *The Pacific Review* 27(3): 333–356.

Hamilton, C. 2013. *Earth Masters. The Dawn of the Age of Climate Engineering*. New Haven and London: Yale University Press.

Hammond, Edward (2009) Indonesia Fights to Change WHO Rules on Flu Vaccines. *Seedling*, April 2009: 24–32.

Hanrieder, Tine (2014) Gradual Change in International Organizations: Agency Theory and Historical Institutionalism. *Politics* 34(4): 324–333.

Hanrieder, Tine (2015) The Path-Dependent Design of International Organizations: Federalism in the World Health Organization. *European Journal of International Relations* 21(1): 215–239.

Harrison, James (2011) *Making the Law of the Sea. A Study in the Development of International Law*. New York: Cambridge University Press.

Harrop, Stuart R. (2011) 'Living in Harmony with Nature'? Outcomes of the 2010 Nagoya Conference of the Convention on Biological Diversity. *Journal of Environmental Law* 23(1): 117–128.

Hart, Sharelle (2008) *Elements of a Possible Implementation Agreement to UNCLOS for the Conservation and Sustainable Use of Marine Biodiversity in Areas Beyond National Jurisdiction*. International Union for the Conservation of Nature, Marine Series No. 4.

Harvey, Alan L., RuAngelie Edrada-Ebel, and Ronald J. Quinn (2015) The Re-emergence of Natural Products for Drug Discovery in the Genomics Era. *Nature Reviews Drug Discovery*, 14(2): 111–29.

Harzevili, Farshad D. (2015) Microbial Biotechnology: An Introduction. In: Farshad D. Harzevili and Hongzhang Chen (eds.), *Microbial Biotechnology. Progress and Trends*. Roca Baton / New York / London: CRC Press.

Hasenclever, Andreas, Peter Mayer, and Volker Rittberger (1996) Interests, Power, Knowledge: the Study of International Regimes. *International Studies Review* 40(2): 177–228.

Hathaway, Oona A. (2001) Path Dependence in the Law: The Course and Patterns of Legal Change in a Common Law System. *Iowa Law Review* 86(2): 101–165.

Heisey, Paul W., C.S. Srinivasan, and Colin Thirstle (2001) *Public Sector Plant Breeding in a Privatizing World*. Washington, DC: US Department of Agriculture.

Helfer, Lawrence R. (2009) Regime Shifting in the International Intellectual Property System. *Perspectives on Politics* 7(1): 39–44.

Helm, Dieter, and Cameron Hepburn (2014) The Economic Analysis of Biodiversity. In: Dieter Helm and Cameron Hepburn (eds.), *Nature in the Balance: The Economics of Biodiversity*. New York: Oxford University Press, 7–32.

Hicks, Bethany L. (1998) Treaty Congestion in International Environmental Law: The Need for Greater International Coordination. *University of Richmond Law Review* 32.

Hoare, Alison L., and Richard G. Tarasofsky (2007) Asking and Telling: Can "Disclosure of Origin" in Patent Applications Make a Difference? *Journal of World Intellectual Property* 10(2): 149–169.

Hodgson, Stephen, Andrew Serdy, Ian Payne, and Johan Gille (2014) *Towards a Possible International Agreement on Marine Biodiversity in Areas Beyond National Jurisdiction*. Brussels: European Parliament.

Howard, Philip H. (2009) Visualizing Consolidation in the Global Seed Industry: 1996–2008. *Sustainability* 1(4): 1266–1287.

Hufty, Marc, Tobias Schulz, and Maurice Tschopp (2014). The Role of Switzerland in the Nagoya Protocol Negotiations. In: Sebastian Oberthür and G. Kristin Rosendal (eds.), *Global Governance of Genetic Resources. Access and Benefit Sharing after the Nagoya Protocol*. Abingdon: Routledge, 96–112.

Hutchison III, Clyde A., et al. (2016) Design and Synthesis of a Minimal Bacterial Genome. *Science* 351: 6280.

ICTSD (2008) Coalition Continues to Press for Demands on Biodiversity and Geographical Indications. International Center for Trade and Sustainable Development. *Bridges* 12(37).

IPW (2005) Consensus Slips on WIPO Patent Harmonisation Talks. *Intellectual Property Watch*, 10 March 2005.

IPW (2008) WHO Members Slow to Bridge Disagreements at Pandemic Flu Meeting. *Intellectual Property Watch*, 11 December 2008.

IPW (2009) Disagreements Remain in WHO Talks on Virus-Sharing after Chan Proposal. *Intellectual Property Watch*, 22 October 2009.

IPW (2011) No Toast Yet to WTO Consensus on Wines and Spirits Geographical Indications. Intellectual Property Watch, 8 November 2011.

Irwin, Rachel (2010) Indonesia, H5N1, and Global Health Diplomacy. *Global Health Governance* 3(2).

Jackson, Patrick T. (2011) *The Conduct of Inquiry in International Relations. Philosophy of Science and Its Implications for the Study of World Politics*. London / New York: Routledge.

Johnson, Tara, and Johannes Urpelainen (2012) A Strategic Theory of Regime Integration and Separation. *International Organization* 66(4): 645–677.

Joly, Pierre-Benoit, and Stéphane Lemarié (1998) Industry Consolidation, Public Attitude and the Future of Plant Biotechnology in Europe. *AgBioForum* 1(2): 85–90.

Josling, Tim (2006) The War on *Terroir*: Geographical Indications as a Transatlantic Trade Conflict. *Journal of Agricultural Economics* 57(3): 337–363.

Joyner, Christopher C. (1986) Legal Implications of the Concept of the Common Heritage of Mankind. *International and Comparative Law Quarterly* 35(1): 190–199.

Juma, Calestous (2011) *The New Harvest. Agricultural Innovation in Africa*. Oxford: Oxford University Press.

Kamradt-Scott, Adam (2012) Changing Perceptions of Pandemic Influenza and Public Health Responses. *American Journal of Public Health* 102(1): 90–98.

Kamradt-Scott, Adam (2015) *Managing Global Health Security. The World Health Organization and Disease Outbreak Control*. Basingstoke: Palgrave.

Kamradt-Scott, Adam, and Kelley Lee (2011) The 2011 Pandemic Influenza Preparedness Framework: Global Health Secured or a Missed Opportunity? *Political Studies* 59(4): 831–847.

Kaul, Inge, Pedro Conceição, Katell Le Goulven, and Ronald U. Mendoza (eds.) (2003) *Providing Global Public Goods. Managing Globalization*. New York / Oxford: Oxford University Press.

Keohane, Robert O., and Joseph S. Nye (2012) *Power and Interdependence*. Boston: Longman.

Keohane, Robert O., and David G. Victor (2011) The Regime Complex for Climate Change. *Perspectives on Politics* 9(1): 7–23.

Kim, Jung-Eun (2013) The Incongruity between the Ecosystem Approach to High Seas Marine Protected Areas and the Existing High Seas Conservation Regime. *Aegean Review of the Law of the Sea and Maritime Law* 2(1): 1–36.

Kinchy, Abby (2012) *Seeds, Science, and Struggle. The Global Politics of Transgenic Crops.* Cambridge, MA / London: MIT Press.

King, Gary, Keohane, Robert O., and Sidney Verba (1994) *Designing Social Inquiry. Scientific Inference in Qualitative Research.* Princeton: Princeton University Press.

King, Gary, and Langche Zeng (2007) When Can History Be Our Guide? The Pitfalls of Counterfactual Inference. *International Studies Quarterly* 51(1): 183–210.

King, John L., Norbert L.W. Wilson, and Anwar Naseem (2002) A Tale of Two Mergers: What Can We Learn from Agricultural Biotechnology Event Studies. *AgBioForum* 5(1): 14–19.

Kleingeld, Pauline (ed.) (2006) *Immanuel Kant. Toward Perpetual Peace and Other Writings on Politics, Peace, and History.* New Haven: Yale University Press.

Kloppenburg, Jack R. (1987) Seed Wars: Common Heritage, Private Property, and Political Strategy. *Socialist Review* 95: 7–41.

Kloppenburg, Jack R. (2004) *First the Seed. The Political Economy of Plant Biotechnology, 1492–2000.* London: University of Wisconsin Press.

Knäblein, Jörg (2013) Twenty Thousand Years of Biotech - From "Traditional" to "Modern" Biotechnology. In: Jörg Knäblein (ed.), *Modern Biopharmaceuticals. Recent Success Stories.* Weinheim: Wiley-Blackwell, 3–38.

Koepsell, David (2015) *Who Owns you? Science, Innovation, and the Gene Patent Wars.* Chichester: Wiley.

Kohler, Pia M., and Melanie Ashton (2010) Paying for POPs: Negotiating the Implementation of the Stockholm Convention in Developing Countries. *International Negotiation* 15(3): 459–484.

Koremenos, Barbara, Charles Lipson, and Duncan Snidal (2001) The Rational Design of International Institutions. *International Organization* 55(4): 761–799.

Koskenniemi, Martti, and Päivi Leino (2002) Fragmentation of International Law? Postmodern Anxieties. *Leiden Journal of International Law* 15: 553–579.

Krasner, Stephen D. (1982) Structural Causes and Regime Consequences: Regimes as Intervening Variables. *International Organization* 36(2): 185–205.

Kur, Annette, and Thomas Dreier (2013) *European Intellectual Property Law. Text, Cases and Materials.* Cheltenham: Edward Elgar.

Laird, Sarah, and Rachel Wynberg (2008) *Access and Benefit-Sharing in Practice: Trends in Partnerships across Sectors.* Convention on Biological Diversity Technical Series No. 38.

Lallas, Peter (2001) The Stockholm Convention on Persistent Organic Pollutants. *American Journal of International Law* 95(3): 692–708.

Lavanchy, Daniel, and Pilar Gavinio (2001) The Importance of Global Influenza Surveillance for the Assessment of the Impact of Influenza. *International Congress Series* 1219: 9–11.

Leary, David K. (2007) *International Law and the Genetic Resources of the Deep Sea.* Leiden / Boston: Martinus Nijhoff Publishers.

Leary, David, Marjo Vierros, Gwenaëlle Hamon, Salvatore Arico, and Catherine Monagle (2009) Marine Genetic Resources: A Review of Scientific and Commercial Interest. *Marine Policy* 33(2): 183–194.

Lebow, Richard N. (2010) *Forbidden Fruit: Counterfactuals and International Relations.* Princeton / Oxford: Princeton University Press.

Leebron, David W. (2002) Linkages. *American Journal of International Law* 96(1): 5–27.

Lenk, Christian (2007) Exclusive Property Rights in the Biosciences: An Ethical Discussion. In: Christian Lenk, Nils Hoppe, and Roberto Andorno (eds.), *Ethics and Law of Intellectual Property. Current Problems in Politics, Science and Technology.* Aldershot: Ashgate, 123–135.

Leskien, Dan (1998) The European Patent Directive on Biotechnology. *Biotechnology and Development Monitor* 36: 16–19.

Li, Jie Jack (2014) *Blockbuster Drugs. The Rise and Decline of the Pharmaceutical Industry.* Oxford / New York: Oxford University Press.

Lightbourne, Muriel (2009) The FAO Multilateral System for Plant Genetic Resources for Food and Agriculture: Better than Bilateralism? *Washington University Journal of Law & Policy* 30: 465–507.

Magulis, Matias E. (2013) The Regime Complex for Food Security: Implications for the Global Hunger Challenge. *Global Governance* 19(1): 53–67.

Mahoney, James (2000) Path Dependence in Historical Sociology. *Theory and Society* 29(4): 507–548.

Mahoney, James, and Kathleen Thelen (2010) A Theory of Gradual Institutional Change. In: James Mahoney and Kathleen Thelen (eds.), *Explaining Institutional Change. Ambiguity, Agency, and Power.* Cambridge: Cambridge University Press.

Mann, Scott (2010) *Bioethics in Perspective: Corporate Power, Public Health and Political Economy.* Cambridge: Cambridge University Press.

Marcoux, Christopher, and Johannes Urpelainen (2012) Capacity, not Constraints: A Theory of North-South Regulatory Cooperation. *Review of International Organizations* 7(4): 399–424.

de Marffy-Mantuano, Annick (1995) The Procedural Framework of the Agreement Implementing the 1982 United Nations Convention on the Law of the Sea. *American Journal of International Law* 89(4): 814–824.

McManis, Charles R. and Eul Soo Seo (2009) The Interface of Open and Proprietary Agricultural Innovation: Facilitated Access and Benefit-Sharing under the New FAO Treaty. *Washington University Journal of Law & Policy* 30: 405–464.

Meunier, Sophie, and Jean-Frédéric Morin (2015) No Agreement Is an Island: Negotiating TTIP in a Dense Regime Complex. In: Sophie Meunier and Jean-Frédéric Morin (eds.), *The Politics of Transatlantic Trade Negotiations. TTIP in a Globalized World.* Aldershot: Ashgate.

Mgbeoji, Ikechi (2006) *Global Biopiracy. Patents, Plants, and Indigenous Knowledge.* Vancouver / Toronto: UBC Press.

Mitchell, Ronald B. (1994) Regime Design Matters: Intentional Oil Pollution and Treaty Compliance. *International Organization* 48(3): 425–458.

Mitchell, Ronald B., and Patricia M. Keilbach (2001) Situation Structure and Institutional Design: Reciprocity, Coercion, and Exchange. *International Organization* 55(4): 891–917.

Mooney, Pat (1983) *The Law of the Seed.* Dag Hammarskjold Foundation, Development Dialogue no.1983: 1–2.

Moore, Gerald, and Emile Frison (2011) International Research Centers: The Consultative Group on International Agricultural Research and the International Treaty. In: Christine Frison, Francisco López and José T. Esquinas-Alcázar (eds.) *Plant Genetic Resources and Food Security. Stakeholder Perspectives on Plant Genetic Resources for Food and Agriculture.* London / New York: Earthscan, 149–162.

Moore, Gerald, and Witold Tymowski (2005) *Explanatory Guide to the International Treaty on Plant Genetic Resources for Food and Agriculture*. Gland / Cambridge: International Union for Conservation of Nature and Natural Resources.

Morgera, Elisa (2007) Competence or Confidence? The Appropriate Forum to Address Multi-Purpose High Seas Protected Areas. *Review of European, Comparative and International Environmental Law* 16(1): 1–11.

Morin, Jean-Frédéric (2009) Multilateralizing TRIPS-plus Agreements: Is the US Strategy a Failure? *Journal of World Intellectual Property* 12(3): 175–197.

Morin, Jean-Frédéric, and Amandine Orsini (2014) Policy Coherence and Regime Complexes: The Case of Genetic Resources. *Review of International Studies* 40(2): 303–324.

Murphy, Denis J. (2007a) *People, Plants, and Genes. The Story of Crops and Humanity*. Oxford / New York: Oxford University Press.

Murphy, Denis J. (2007b) *Plant Breeding and Biotechnology. Societal Context and the Future of Agriculture*. Cambridge: Cambridge University Press.

Murray, Fiona, and Scott Stern (2007) Do Formal Intellectual Property Rights Hinder the Free Flow of Information? An Empirical Test of the Anti-Commons Hypothesis. *Journal of Economic Behavior & Organization* 63(4): 648–687.

Muzaka, Valbona (2010) Linkages, Contests and Overlaps in the Global Intellectual Property Rights Regime. *European Journal of International Relations* 17(4): 755–776.

Muzaka, Valbona (2011) *The Politics of Intellectual Property Rights and Access to Medicines*. Basingstoke: Palgrave Macmillan.

Mwila, Godfrey (2013) From Negotiations to Implementation: Global Review of Achievements, Bottlenecks and Opportunities for the Treaty in General and for the Multilateral System in Particular. In: Michael Halewood, Isabel López Noriega, and Selim Louafi (eds.), *Crop Genetic Resources as a Global Commons: Challenges in International Law and Governance*. London: Earthscan, 226–242.

NAS (2009) *Financing Vaccines in the 21st Century: Assuring Access and Availability*. Washington, DC: National Academies Press.

NAS (2016a) *Gene Drives on the Horizon. Advancing Science, Navigating Uncertainty, and Aligning Research with Public Values*. Washington, DC: National Academies Press.

NAS (2016b) *Genetically Engineered Crops: Experiences and Prospects*. Washington, DC: National Academies Press.

Nightingale, Paul, and Surya Mahdi (2006) The Evolution of Pharmaceutical Innovation. In: Mariana Mazzucato and Giovanni Dosi (eds.), *Knowledge Accumulation and Industry Evolution. The Case of Pharma-Biotech*. Cambridge: Cambridge University Press, 73–111.

Nijar, Gurdial S. (2011) *The Nagoya Protocol on Access and Benefit Sharing of Genetic Resources: An Analysis*. Kuala Lumpur: Centre of Excellence for Biodiversity Law (CEBLAW).

Ninan, K.N. (ed.) (2009) *Conserving and Valuing Ecosystem Services and Biodiversity. Economic, Institutional and Social Challenges*. London / Sterling, VA: Earthscan.

van Notten, Philip W. F., Jan Rotmans, Marjolein B.A. van Asselt, and Dale S. Rothman (2003) An Updated Scenario Typology. *Futures* 35(5): 423–443.

Oberthür, Sebastian (2009) Interplay Management: Enhancing Environmental Policy Integration among International Institutions. *International Environmental Agreements* 9(4): 371–391.

Oberthür, Sebastian, and Justyna Pozarowska (2013) Managing Institutional Complexity and Fragmentation: The Nagoya Protocol and the Global Governance of Genetic Resources. *Global Environmental Politics* 13(3): 100–118.

Oberthür, Sebastian, and Florian Rabitz (2014) On the EU's Performance and Leadership in Global Environmental Governance: The Case of the Nagoya Protocol. *Journal of European Public Policy* 21(1): 39–57.

Oberthür, Sebastian, and Olav S. Stokke (2011) Conclusions: Decentralized Interplay Management in an Evolving Interinstitutional Order. In: Sebastian Oberthür and Olav S. Stokke (eds.), *Managing Institutional Complexity. Regime Interplay and Global Environmental Change.* Cambridge, MA: MIT Press, 313–342.

OECD (2005) *A Framework for Biotechnology Statistics.* Paris: Organization for Economic Co-operation and Development.

OECD (2011) *Future Prospects for Industrial Biotechnology.* Paris: Organization for Economic Co-operation and Development.

OECD (2013) *Marine Biotechnology: Enabling Solutions for Ocean Productivity and Sustainability.* Paris: Organization for Economic Co-operation and Development.

Olson, David M., Eric Dinerstein, and 16 others (2001) Terrestrial Ecoregions of the World: A New Map of Life on Earth. *BioScience* 51(11): 933–938.

Olson, Mancur (1975) *The Logic of Collective Action. Public Goods and the Theory of Groups.* Cambridge, MA / London: Harvard University Press.

Orsini, Amandine (2013) Multi-Forum Non-State Actors: Navigating the Regime Complexes for Forestry and Genetic Resources. *Global Environmental Politics* 13(3): 34–55.

Orsini, Amandine, Jean-Frédéric Morin, and Oran R. Young (2013) Regime Complexes: A Buzz, a Boom, or a Boost for Global Governance? *Global Governance: A Review of Multilateralism and International Organizations* 19(1): 27–39.

Oyelaran-Oyeyinka, Banji, and Padmashree Gehl Sampath (2009) *The Gene Revolution and Global Food Security. Biotechnology Innovation in Latecomers.* Basingstoke: Palgrave Macmillan.

Palombi, Luigi (2009) *Gene Cartels. Biotech Patents in the Age of Free Trade.* Cheltenham: Edward Elgar.

Pandey, Ashok, Rainer Höfer, Mohammad Taherzadeh, K. Madhavan Nampoothiri, and Christian Larroche (eds.) (2015) *Industrial Biorefineries and White Biotechnology.* Amsterdam: Elsevier.

Pardey, Philip G., Julian M. Alston, and Connie Chan-Kang (2013) Public Agricultural R&D over the Past Half Century: An Emerging New World Order. *Agricultural Economics* 44: 103–113.

Parisi, Claudia, Pascal Tillie, and Emilio Rodriguez-Cerezo (2016) The Global Pipeline of GM Crops out to 2020. *Nature Biotechnology* 34(1): 31–36.

Pearce, David, and Seema Puroshothaman (1995) The Economic Value of Plant-Based Pharmaceuticals. In: Timothy Swanson (ed.), *Intellectual Property Rights and Biodiversity Conservation: An Interdisciplinary Analysis of the Values of Medicinal Plants.* Cambridge: Cambridge University Press, 127–138.

Petit, Michel, Cary Fowler, Wanda Collins, Carlos Correa, and Carl-Gustaf Thomström (2001) *Why Governments Can't Make Policy. The Case of Plant Genetic Resources in the International Arena.* Lima: International Potato Center.

PhRMA (2013) *Vaccine Fact Book 2013.* Washington, DC: Pharmaceutical Research and Manufacturers of America.

Pierson, Paul (2004) *Politics in Time: History, Institutions and Social Analysis.* Princeton: Princeton University Press.

Presidential Commission (2010) *New Directions. The Ethics of Synthetic Biology and Emerging Technologies.* Washington, DC: Presidential Commission for the Study of Bioethical Issues.

Price-Smith, Andrew T. (2009) *Contagion and Chaos. Disease, Ecology, and National Security in the Era of Globalization.* Cambridge, MA: MIT Press.

Pujar, Narahari S., Sangeetha L. Sagar, and Ann L. Lee (2015) History of Vaccine Process Development. In: Emily P. Wen, Ronald Ellis, and Narahari S. Pujar (eds.), *Vaccine Development and Manufacturing.* Hoboken: Wiley, 1–24.

Qaim, Matin, and David Zilberman (2003) Yield Effects of Genetically Modified Crops in Developing Countries. *Science* 299: 900–902.

Rabitz, Florian (2014) Explaining Institutional Change in International Patent Politics. *Third World Quarterly* 35(9): 1582–1597.

Rabitz, Florian (2015) Biopiracy after the Nagoya Protocol: Problem Structure, Regime Design and Implementation Challenges. *Brazilian Political Science Review* 9(2): 30–53.

RAFI (2001) *Enclosures of the Mind: Intellectual Monopolies.* Ottawa: Rural Advancement Foundation International.

Rasmussen, Bruce (2010) *Innovation and Commercialization in the Biopharmaceutical Industry. Creating and Capturing Value.* Cheltenham: Edward Elgar.

Rasmussen, Nicolas (2014) *Gene Jockeys. Life Sciences and the Rise of the Biotech Enterprise.* Baltimore: Johns Hopkins University Press.

Raustiala, Kal, and David G. Victor (2004) The Regime Complex for Plant Genetic Resources. *International Organization* 58(2): 277–309.

Renkow, Mitch (2010) The Impacts of CGIAR Research. *Food Policy* 35(5): 391–402.

Richardson, Ben (2012) From a Fossil-Fuel to a Biobased Economy: The Politics of Industrial Biotechnology. *Environment and Planning C: Government and Policy* 30(2): 282–296.

Rimmer, Matthew (2008) *Intellectual Property and Biotechnology. Biological Inventions.* Cheltenham: Edward Elgar.

Robinson, Daniel F. (2010) *Confronting Biopiracy. Challenges, Cases and International Debates.* London: Earthscan.

Rochette, Julien, Kristina Gjerde, et al. (2014) Delivering the Aichi Target 11: Challenges and Opportunities for Marine Areas Beyond National Jurisdiction. *Aquatic Conservation: Marine and Freshwater Ecosystems* 24(s.2): 31–43.

Rochette, Julien, Glen Wright, Kristina M. Gjerde, Thomas Greiber, Sebastian Unger, and Aurélie Spadone (2015) A New Chapter for the High Seas? Historic Decision to Negotiate an Internationally Legally Binding Instrument on the Conservation and Sustainable Use of Marine Biodiversity in Areas Beyond National Jurisdiction. *IASS Working Paper.* Potsdam. Institute for Advanced Sustainability Studies.

de Rond, Mark (2003) *Strategic Alliances as Social Facts. Business, Biotechnology, and Intellectual History.* Cambridge: Cambridge University Press.

Rosendal, Kristin (2000) *The Convention on Biological Diversity and Developing Countries.* Boston: Kluwer Academics.

Rosendal, Kristin (2006) The Convention on Biological Diversity: Tensions with the WTO TRIPS Agreement over Access to Genetic Resources and the Sharing of Benefits. In: Sebastian Oberthür and Thomas Gehring (eds.), *Institutional Interaction in Global Environmental Governance. Synergy and Conflict among International and EU Policies.* Cambridge, MA: MIT Press, 79–102.

Rosendal, Kristin, and Steinar Andresen (2014) Realizing Access and Benefit-Sharing from Use of Genetic Resources Between Diverging International Regimes: The Scope for Leadership. *International Environmental Agreements*, online first. doi: 10.1007/s10784-014-9271-4.

Sampath, Padmashree Gehl (2005) *Regulating Bioprospecting: Institutions for Drug Research, Access, and Benefit-Sharing.* Tokyo / New York / Paris: United Nations University.

Schaffrin, Dora, Benjamin Görlach, and Christiane Gerstetter (2006) *The International Treaty on Plant Genetic Resources for Food and Agriculture – Implications for Developing Countries and Interdependence with International Biodiversity and Intellectual Property Law.* Berlin: Ecologic Institute.

Schelling, Thomas (1973) Hockey Helmets, Concealed Weapons, and Daylight Saving: A Study of Binary Choices with Externalities. *Journal of Conflict Resolution* 17(3): 381–428.

Schimmelfennig, Frank (2015) Efficient Process Tracing: Analyzing the Causal Mechanisms of European Integration. In: Andrew Bennett and Jeffrey T. Checkel (eds.), *Process Tracing. From Metaphor to Analytic Tool.* Cambridge: Cambridge University Press, 98–125.

Schweizer, Lars (2005) Organizational Integration of Acquired Biotechnology Companies into Pharmaceutical Companies: The Need for a Hybrid Approach. *Academy of Management Journal* 48(6): 1051–1074.

Sebenius, James K. (1983) Negotiation Arithmetic: Adding and Subtracting Issues and Parties. *International Organization* 37(2): 281–316.

Sedyaningsih, E.R., S. Isfandari, T. Soendoro, and S.F. Supari. (2008) Towards Mutual Trust, Transparency and Equity in Virus Sharing Mechanism. The Avian Influenza Case of Indonesia [sic]. *Annals of the Medical Academy of Singapore* 37: 482–488.

Selin, Henrik, and Noelle Eckley (2003) Science, Politics and Persistent Organic Pollutants. The Role of Scientific Assessments in International Environmental Co-operation. *International Environmental Agreements* 3(1): 17–42.

Sell, Susan (2010) TRIPS Was Never Enough: Vertical Forum Shifting, FTAs, ACTA, and TPP. *Journal of Intellectual Property Law* 18(2): 447–478.

Shah, Sonia (2016) *Pandemic. Tracking Contagions, From Cholera and Ebola and Beyond.* New York: Farrar, Straus and Giroux.

Shapiro, Carl (2001) Navigating the Patent Thicket: Cross Licenses, Patent Pools, and Standard Setting. In: Adam B. Jaffe, Josh Lerner, and Scott Stern (eds.), *Innovation Policy and the Economy, Vol. I.* Cambridge, MA: MIT Press, pp. 119–150.

Shiva, Vandana (1993) *The Violence of the Green Revolution. Third World Agriculture, Ecology and Politics.* London: Zed Books.

Shiva, Vandana (2001) *Protect or Plunder? Understanding Intellectual Property Rights.* London: Zed Books.

Sivamani, Raja K., Jared R. Jagdeo, Peter Elsner, and Howard I. Maibach (eds.) (2016) *Cosmeceuticals and Active Cosmetics.* Boca Raton / London / New York: CRC Press.

Smallman, S. (2013) Biopiracy and Vaccines: Indonesia and the World Health Organization's New Pandemic Influenza Plan. *Journal of International & Global Studies* 4(2): 20–36.

Smith III, Frank L (2015) Insights into Surveillance from the Influenza Virus and Benefit Sharing Controversy. In: Sara E. Davies and Jeremy R. Youde (eds.), *The Politics of Surveillance and Response to Disease Outbreaks.* Aldershot: Ashgate, 212–136.

Smith Hughes, Sally (2012) *Genentech: The Beginnings of Biotech.* Chicago / London: University of Chicago Press.

Snidal, Duncan (1985) The Limits of Hegemonic Stability Theory. *International Organization* 39(4): 579–614.

Soetaert, Wim, and Erick Vandamme (2005) The Impact of Industrial Biotechnology. *Biotechnology Journal* 1(7–8): 756–769.

Srinisavan, C. S. (2010) Plant Breeders' Rights and On-farm Seed-Saving. In: Stewart Lockie and David Carpenter (eds.), *Agriculture, Biodiversity and Markets. Livelihoods and Agroecology in Comparative Perspective.* London / Washington, DC: Earthscan, 61–76.

Stannard, Clive (2013) The Multilateral System of Access and Benefit Sharing: Could It Have Been Constructed Another Way? In: Michael Halewood, Isabel López Noriega and Selim Louafi (eds.), *Crop Genetic Resources as a Global Commons: Challenges in International Law and Governance.* London: Earthscan, 243–264.

Stöhr, Klaus, and Marja Esveld (2004) Will Vaccines be Available for the Next Influenza Pandemic? *Science* 306(5705): 2195–2196.

Stokke, Olav S. (2013) Regime Interplay in Arctic Shipping Governance: Explaining Regional Niche Selection. *International Environmental Agreements* 13(1): 65–85.

Stoll, Peter-Tobias (2009) Access to GRs and Benefit Sharing – Underlying Concepts and the Idea of Justice. In: Evanson C. Kamau and Gerd Winter (eds.), *Genetic Resources, Traditional Knowledge and the Law. Solutions for Access and Benefit Sharing.* London: Earthscan, 3–18.

Streeck, Wolfgang, and Kathleen Thelen (2005) Introduction: Institutional Change in Advanced Political Economies. In: Wolfgang Streeck and Kathleen Thelen (eds.), *Beyond Continuity: Institutional Change in Advanced Political Economies.* Oxford: Oxford University Press, 1–39.

Struett, Michael J., Mark T. Nance, and Diane Armstrong (2013) Navigating the Maritime Piracy Regime Complex. *Global Governance* 19(1): 93–104.

SUNS (2007) WHO Meeting on Avian Influenza Virus Ends with Draft Documents. South-North Development Bulletin. Geneva: Third World Network.

Sylvan, David, and Stephen Majeski (1998) A Methodology for the Study of Historical Counterfactuals. *International Studies Quarterly* 42(1): 79–108.

Tan, Alan Khee-Jin (2005) *Vessel-Source Marine Pollution. The Law and Politics of International Regulation.* Cambridge: Cambridge University Press.

Tanaka, Yoshifumi (2012) *The International Law of the Sea.* New York: Cambridge University Press.

Ten Kate, Kerry, and Carolina Lasén Diaz (1997) The Undertaking Revisited: A Commentary on the Revision of the International Undertaking on Plant Genetic Resources for Food and Agriculture. *Review of Comparative and International Environmental Law* 6(3): 284–292.

Ten Kate, Kerry, and Sarah A. Laird (2000) *The Commercial Use of Biodiversity: Access to Genetic Resources and Benefit-Sharing.* London / Sterling, VA: Earthscan.

Thelen, Kathleen (2004) *How Institutions Evolve. The Political Economy of Skills in Germany, Britain, the United States, and Japan.* Cambridge: Cambridge University Press.

Tripp, Robert, Niels Louwaars, and Derek Eaton (2007) Plant Variety Protection in Developing Countries. A Report from the Field. *Food Policy* 32(3): 354–371.

Tully, Stephen (2003) The Bonn Guidelines on Access to Genetic Resources and Benefit Sharing. *Review of European, Comparative and International Environmental Law* 12(1): 84–98.

Turner, Mark (2015) Vaccine Procurement during an Influenza Pandemic and the Role of Advance Purchase Agreements: Lessons from 2009-H1N1. *Global Public Health*, doi: g/10.1080/17441692.2015.1043743.

Tvedt, Morten Walløe (2013) Bioprospecting in the High Seas: Regulatory Options for Benefit Sharing. *Journal of World Intellectual Property* 16(3–4): 150–167.

Tvedt, Morten Walløe, and Peter Johan Schei (2014) The Term 'Genetic Resources': Flexible and Dynamic while Providing Legal Certainty? In: Sebastian Oberthür and G. Kristin Rosendal (eds.), *Global Governance of Genetic Resources. Access and Benefit Sharing after the Nagoya Protocol*. Abingdon: Routledge, 18–32.

TWN (2007) No Clear Outcomes in WHO Meeting on Avian Flu Virus Sharing. Third World Network.

TWN (2009) Key Elements of Virus and Benefit-Sharing Framework Still Unresolved. SUNS #6703, 19 May 2009.

TWN (2010) WHO: Virus-Benefit Sharing Working Group Set Up, Pandemic Flu Response to be Reviewed. SUNS #6849, 26 January 2010.

Urpelainen, Johannes, and Thijs van de Graaf (2015) Your Place or Mine? Institutional Capture and the Creation of Overlapping International Institutions. *British Journal of Political Science* 45(4): 799–827.

van Asselt, Harro (2014) *The Fragmentation of Global Climate Governance. Consequences and Management of Regime Interactions*. Cheltenham: Edward Elgar.

van Beuzekom, Brigitte, and Anthony Arundel (2009) *OECD Biotechnology Statistics 2009*. Paris: Organization for Economic Co-operation and Development.

van den Hurk, Anke (2011) The Seed Industry: Plant Breeding and the International Treaty on Plant Genetic Resources for Food and Agriculture. In: Christine Frison, Francisco López, and José T. Esquinas-Alcázar (eds.) *Plant Genetic Resources and Food Security. Stakeholder Perspectives on Plant Genetic Resources for Food and Agriculture*. London / New York: Earthscan, 163–174.

van der Heijden, Jeroen (2011) Institutional Layering: A Review of the Use of the Concept. *Politics* 31(1): 9–18.

Vezzani, Simone (2010) Preliminary Remarks on the Envisaged World Health Organization Pandemic Influenza Preparedness Framework for the Sharing of Viruses and Access to Vaccines and other Benefits. *Journal of World Intellectual Property* 13(6): 675–696.

Victor, David G. (2011) *Global Warming Gridlock. Creating More Effective Strategies for Protecting the Planet*. Cambridge: Cambridge University Press.

Visser, Bert, and Jan Borrin (2011) The European Regional Group: Europe's Role and Position during the Negotiations and Early Implementation of the International Treaty. In: Christine Frison, Francisco López and José T. Esquinas-Alcázar (eds.) *Plant Genetic Resources and Food Security. Stakeholder Perspectives on Plant Genetic Resources for Food and Agriculture*. London / New York: Earthscan, 69–80.

Vivas-Eugui, David (2012) *Bridging the Gap on Intellectual Property and Genetic Resources in WIPO's Intergovernmental Committee (IGC)*. Geneva: International Center for Trade and Sustainable Development.

Wallbott, Linda (2014) Goals, Strategies and Success of the African Group in the negotiations of the Nagoya Protocol. In: Sebastian Oberthür and Kristin Rosendal (eds.), *Global Governance of Genetic Resources. Access and Benefit Sharing after the Nagoya Protocol*. New York / London: Routledge.

Wallbott, Linda, Franziska Wolff, and Justyna Pozarowska (2014) The Negotiations of the Nagoya Protocol. Issues, Coalitions and Process. In: Sebastian Oberthür and Kristin Rosendal (eds.), *Global Governance of Genetic Resources. Access and Benefit Sharing after the Nagoya Protocol*. New York / London: Routledge.

Wallerstein, Immanuel (1983) *Historical Capitalism*. London: Verso.

Warzecha, Heribert (2012) Biopharmaceuticals from Plants. In: Oliver Kayser and Heribert Warzecha (eds.), *Pharmaceutical Biotechnology.* Weinheim: Wiley-Blackwell, pp. 59–69.

WIPO (2011) *WIPO Patent Search Report on Pandemic Influenza Preparedness (PIP)-related Patents and Patent Applications.* Geneva: World Intellectual Property Organization.

Wynberg, Rachel, Doris Schroeder, and Roger Chennells (eds.) (2009) *Indigenous Peoples, Consent and Benefit Sharing: Lessons from the San-Hoodia Case.* Dordrecht: Springer.

Wynne, Brian (1989) The Toxic Waste Trade: International Regulatory Options and Issues. *Third World Quarterly* 11(3): 120–146.

Young, Oran (1991) Political Leadership and Regime Formation: On the Development of Institutions in International Society. *International Organization* 45(3): 281–308.

Young, Oran (1996) Institutional Linkages in International Society: Polar Perspectives. *Global Governance* 2(1): 1–24.

Zelli, Fariborz, and Harro van Asselt (2013) Introduction: The Institutional Fragmentation of Global Environmental Governance: Causes, Consequences, and Responses. *Global Environmental Politics* 13(3): 1–13.

Zürn, Michael (1992) *Interessen und Institutionen in der internationalen Politik. Grundlegung und Anwendung des situationsstrukturellen Ansatzes.* Springer: Wiesbaden.

Index